A LIFE OF
MAGIC CHEMISTRY

A LIFE OF MAGIC CHEMISTRY

Autobiographical Reflections
Including Post-Nobel Prize Years
and the Methanol Economy

Second Updated Edition

George A. Olah

with

Thomas Mathew

WILEY

Published by John Wiley & Sons, Inc., Hoboken, New Jersey
Published simultaneously in Canada

For general information on our other products and services or for technical support, please contact our Customer Care Department within the United States at (800) 762-2974, outside the United States at (317) 572-3993 or fax (317) 572-4002.

Wiley also publishes its books in a variety of electronic formats. Some content that appears in print may not be available in electronic formats. For more information about Wiley products, visit our web site at www.wiley.com.

Library of Congress Cataloging-in-Publication Data:

Olah, George A. (George Andrew), 1927–
    A life of magic chemistry : autobiographical reflections including post-Nobel Prize years and the methanol economy / George A. Olah with Thomas Mathew. – Second updated edition.
        pages cm
    Includes bibliographical references and index.
    ISBN 978-1-118-84003-0 (cloth)
    1. Olah, George A. (George Andrew), 1927–   2. Chemists–United States–Biography.
I. Mathew, Thomas, 1961–   II. Title.
    QD22.O43A3 2015
    540.92–dc23
    [B]   2015005387

Cover – Oil portrait by Kornél Zámbó

Printed in the United States of America

10 9 8 7 6 5 4 3 2 1

To Judy,
who made it all possible

My grandchildren Kaitlyn, Justin and Peter (December 2013)

Contents

Preface

My wife Judy and friends urged me to update my original autobiography published in 2001 by Wiley. In this second edition, I included much of my original autobiographical recollections of my life and my work in chemistry, adding particularly those of the intervening years. But I also added some of my more general thoughts on science. These touch on topics including the broader meaning of science in the quest for understanding and knowledge as well as their limitations. Science as a human endeavor means the search for knowledge about the physical world and the natural sciences governing it (both physical and life sciences). It does not include the spiritual or supernatural (including the area of philosophy). Inevitably, however, this leads to such fundamental questions of how it all started and developed: Was there a beginning? Was our being planned by a higher intelligence? It is an exciting struggle with these and related questions while trying to balance what we know through science and what we must admit are beyond us. My thoughts are those of a scientist who always tried, however imperfectly, to maintain his early interest in the classics, history, philosophy, and the arts. In recent years, I have tried increasingly to fill in some of the gaps, but a life actively pursuing science inevitably imposes constraints on the time that one can spend reading and studying outside one's own field of specialization. Of course, I realize only too well my limitations and the lack of depth of my background in some of these areas. Therefore, I have tried not to overreach, and I will limit my thoughts to my own understanding and views, however imperfect they may be.

This book is mainly about my life in search of chemistry. Because some of my work centered on the discovery and use of extremely strong "superacids," some of which are now also called "magic acids," I choose again the title *A Life of Magic Chemistry*. It also reflects, however, in a more general way, on the exciting and sometimes indeed even "magic" nature of chemistry, which with its extremely broad scope cuts through many of the varied sciences, truly being a central one. In recent years, I concentrated my interest to find new solutions to our depleting fossil fuel resources, particularly substituting oil with methanol in the scope of the "Methanol Economy."

It was a long journey that led me from the banks of the Danube in Budapest, through Cleveland, to the shore of the Pacific Ocean in Los Angeles, with a side trip to Stockholm. Sometimes I still wonder how life unfolds in ways we could not have planned or even foreseen.

I thank my publisher for the interest in publishing, after nearly 15 years, an updated second edition of my original autobiography. Ms. Anita Lekhwani and Cecilia Tsai have rendered their help to get the updated second edition in print. My wife, sons, friends, and colleagues helped to improve the manuscript and commented on its many shortcomings. My particular thanks go to Jessie May, my administrative assistant, who helped in the preparation of the manuscripts and progress of the book. But mainly, I thank my colleague and friend Dr. Thomas Mathew, who through his outstanding effort made the second updated and extended edition possible.

George A. Olah
Los Angeles
July 2014

· 1 ·

Introduction

If we look back on the history of human efforts for understanding of our world and of the universe, these look like lofty goals that, I believe, mankind will never fully achieve. In earlier times, things were more simple. The great Greek thinkers and those who followed in their footsteps were able to combine the knowledge available of the physical world with their thoughts of the "spiritual world" and thus develop their overall philosophy. This changed with the expansion of scientific inquiry and quest for knowledge in the seventeenth century. By the twentieth century, few philosophers, except those who also had some background in the sciences, could claim sufficient knowledge of the physical world to even attempt serious consideration of its meaning. This opened up for some scientists, particularly physicists, the claim to center stage, suggesting that only science can attempt to give answers to such fundamental questions as the origin and meaning of the universe, life, our being as intelligent species, and the understanding of the universal laws governing the physical and biological world. In reality, however, humankind with all its striving for such knowledge probably never will reach full understanding. For me this is readily acceptable. It seems only honest to admit our limitations because of which human knowledge can reach only a certain point. Our knowledge will continue to expand, but it hardly can be expected to give answers to many of the fundamental questions of mankind. Chemists do not need to claim fundamental insights in the ways in which the atoms of the elements were formed after

A Life of Magic Chemistry: Autobiographical Reflections Including Post-Nobel Prize Years and the Methanol Economy, Second Updated Edition. George A. Olah with Thomas Mathew.

the initial "big bang," because they are concerned only with their eventual assembly into molecules (compounds, materials). They can avoid the question of whether all these were planned and created with a predetermined goal. I will, however, briefly reflect on my own views and thoughts. They reflect to the struggle and inevitable compromises, leading to what I consider—at least for me—an acceptable overall realization that we, in all probability, never can expect a full understanding.

Science is humankind quest for better understanding and discovery of the physical and biological (life) aspects of what exists. Pasteur, the great French scientist of his time wrote, "There does not exist a category of science to which one can give the name of applied science." This certainly is valid to the multitude of the numerous hyphenated sciences of our present time. For example, the important and significant areas (of social and political studies) and many what I would call "hyphenated sciences" which are to some extent scientific methods. They do not represent however a separate applied branch of science. They are only applications of the findings of essential basic science to the practical (applied) areas to the benefit of humankind. Applied sciences are only applications of essential knowledge of fundamental science to practical areas. They are not a specific separate area of science, which are so generally supported and pursued by society (through private individuals, governments, political parties, and public opinions). Basic (fundamental) and applied science are both essential of our never ending quest for knowledge and its application.

I was lucky to be able to work during and contribute to one of the most exciting period of science, that of the second half of the twentieth century and the beginning of the twenty-first century. I was also fortunate that I was mostly able to pursue my own interests in chemistry, following my own way and crossing conventional boundaries. Frequently, I left behind what Thomas Kuhn called safe, "normal science" in pursuit of more exciting, elusive new vistas. How many people can say that they had a fulfilling, happy life doing what they love to do and were even paid for it? Thus, when people ask me whether I still work, my answer is that I do, but chemistry was never really work for me. It was and still is my passion, my hobby. I do not have many other interests outside chemistry, except for my family and my continuous learning about a wide range of topics through reading. Thus the long hours I still spend on science come naturally to me and are very enjoyable. If, one day, the joy and satisfaction that chemistry gives me should cease or my capabilities decline so that I can make no further meaningful contributions (including helping my younger colleagues in their own development and efforts), I will walk away from it without hesitation.

I always was interested in attempting to link the results of my basic research with practical uses including environmentally friendly ways. This in recent years meant finding new ways of producing hydrocarbon fuels and derived materials and chemicals that at the same time also safeguard our fragile environment. With my colleague and friend Surya Prakash and our colleagues in our Institute we developed the concept and much enabling new chemistry of what now is called "The Methanol Economy" which is gaining worldwide practical application. Pinpointing environmental and health hazards and then eliminating them are another part of our efforts. It is through finding new solutions and answers to global problems that we can work for a better future. In this regard chemistry can offer much. I find it extremely rewarding that my colleagues and I can increasingly contribute to these practical goals through our work. This also shows that there is no dichotomy between gaining new knowledge through basic research and finding practical uses for it. It is the most rewarding aspect of chemistry that in many ways it can not only contribute to our better understanding of the physical and biological world but also supplement nature by allowing man to produce through his own efforts essential solutions, products, and materials to allow future generations a better life while also protecting our environment.

· 2 ·

Perspectives on Science

I have spent my life in science pursuing the magic of chemistry. In attempting to give some perspectives and thoughts on science, it is first necessary to define what science really is. As with other frequently used (or misused) terms such as "God" or "democracy" that have widely differing meanings to different people at different times and places "science" does not seem to be readily and uniformly defined. Science, derived from the Latin "scientia," originally meant general knowledge both of the physical and spiritual world. Through the ages, however, the meaning of science narrowed to the description and understanding (knowledge) of nature (i.e., the physical world). Science is thus a major intellectual effort of man, a search for knowledge of the physical world, the laws governing it, and its meaning. It also touches however on fundamental, ageless questions as to our existence, origin, purpose, intelligence, etc., and, through these, the limits of how far our understanding can reach. In many ways the scientists' intellectual efforts expressing their thoughts and quest for general knowledge and understanding are similar to other intellectual efforts in areas such as the humanities, arts, etc., although they are expressed in different ways.

In discussing science we also need to define its scope, as well as the methods and views (concepts) involved in its pursuit. It is also useful to think about what science is *not*, although this can sometimes become controversial. Significant and important studies such as those concerned

A Life of Magic Chemistry: Autobiographical Reflections Including Post-Nobel Prize Years and the Methanol Economy, Second Updated Edition. George A. Olah with Thomas Mathew.
© 2015 John Wiley & Sons, Inc. Published 2015 by John Wiley & Sons, Inc.

with the fields of sociology, politics, or economics increasingly use methods that previously were associated only with the physical and biological sciences or mathematics. However, I believe these are not in a strict sense "hard sciences." The name "science" these days is also frequently hyphenated to vary other fields (from animal-science to culinary science to exercise science, etc.). Such studies indeed may use some of the methods of science, but they hardly fall under the scope of science. There is a Dutch proverb that says "Everything has its science, with the exception of catching fleas: This is an art." It may overstate the point, but sometimes to make a point it is necessary to overstate it.

When we talk about *knowledge* of the physical world we generally refer to facts derived from systematic observation, study, and experimentation as well as the concepts and theories based on these facts. This is contrasted with *belief* (faith, intuition) of the spiritual or supernatural.

Scientists use methods in their pursuit of knowledge that frequently are referred to collectively as the "scientific method." Originally credit goes to Francis Bacon's ideas at the end of the sixteenth century. Bacon believed that the facts in any given field can be collected according to accepted and prearranged plans and then passed through a logical intellectual process from which the correct judgments will emerge. Because phenomena (facts) were so numerous even then, he suggested that they must be chosen (selected), which is a subjective act of judgment. This process is however hardly compatible with what we now associate with the scientific method.

This also brings up the essential relationship of science and its historical perspective. We can never talk about science, without putting it into its time frame. August Comte wrote, "L'histoire de la science c'est la science meme" – "The history of science is science itself." For example, when we look back in time early scientists (savants) long believed that the earth is the center of the universe and that it is flat. They even warned that approaching its edges would put one at risk of falling off. However strange this may seem to be for us today, they were interpreting the limited knowledge they had at the time. We may pride ourselves on what we consider our advanced knowledge in the twenty-first century, but I am sure future generations will look back at us and say how ignorant and naive we were. As Einstein said, "One thing I have learned in a long life is that all of our science, measured against reality, is primitive and child-like and yet it is the most precious thing we have." I hope that it will also be remembered that we tried our best. Scientific knowledge by its nature continuously changes and expands. Only through its historical time frame can science be put into its proper perspective. It is thus regrettable that the history of science is generally not taught in our universities and colleges. This probably is also due to the fact that the interactions between scientists and historians (philosophers), and the mutual understanding of the significance of their fields, are frequently far from satisfactory.

The days are long gone when friends of Lavoisier, one of the greatest scientists of the time, during the terror of the French revolution, were pleading for his life before the revolutionary tribunal. It, however, ruled that "la revolution n'a pas besoin de la science" (the revolution does not need science). His head was guillotined off the same day. Since that time it has however become clear that the world needs very much science for a better future. Science does not know national, racial, or religious distinctions. There is no separate American, European, Chinese, or Indian science; science is truly international. Although scientific results, like anything else, can also be misused (the study of atomic energy is still frequently condemned because its development was closely related to that of the atom bomb), we cannot be limited and must look at the broader benefits of science.

The scientific method, as mentioned before, involves observation and experimentation (research) undertaken to discover or establish facts. These are followed by deduction or hypothesis, establishing theories or principles. This sequence, however, may be reversed. The noted twentieth century philosopher Karl Popper, who also dealt with science, expressed the view that the scientist's work starts not with collection of data (observation) but with selection of a suitable problem (theory). In fact, however, both of these paths can be involved. Significant and sometimes accidental observations can be made without any preconceived idea of a problem or theory and vice versa. The scientist, however, must have a well-prepared, open mind to be able to recognize the significance of such observations and must be able to follow them through. Science always demands rigorous standards of procedure, reproducibility, and open discussion that sets reason over irrational belief.

Research is frequently considered to be either basic (to build up fundamental knowledge) or applied (directed to specific practical goals). I myself, however, have never believed in such strict dividing line. Whenever I made some new finding in chemistry I never could resist also exploring whether it might have a practical use. The results of scientific research can subsequently be developed into technology (research and development). It is necessary however to differentiate science from technology, because they are frequently lumped together without clearly defining their differences. To recapitulate: *science* is the search for knowledge; *technology* is the application of scientific knowledge to provide for the needs of society (in a practical as well as economically feasible way).

"In the pursuit of research or observation many would see what others have seen before, but it is the well-prepared one who (according to Albert Szent-Györgyi, Nobel Prize winner in medicine 1937) may think what nobody else has thought before" and achieve a discovery or breakthrough. Mark Twain once wrote that "the greatest of all inventors is chance." Chance, however, will favor only those who are prepared for it and capable of recognizing the significance of an unexpected invention and explore it further.

Thomas Kuhn, the science philosopher, in his *Structure of Scientific Revolutions*, called "normal science" research that is based upon established and accepted concepts (paradigms) that are acknowledged as providing the foundation for the future. This is the overwhelming part of scientific research. It is also considered "safe" to pursue because it is rarely controversial. Following Yogi Berra's advice, it allows the scientists "not to make the wrong mistakes." Consequently, it is usually supported and peer approved. Some scientists, however, point out occasionally unexpected and unexplained new findings or observed anomalies. These always are "high risk" and controversial and frequently turn out to be wrong. But on occasions they can lead to new fundamental scientific discoveries and breakthroughs that advance science to new levels (paradigm changes). Kuhn called this "revolutionary science," which can lead to groundbreaking discoveries which cannot be accommodated by existing paradigms.

Science continuously develops ever more rigorous standards of procedure and evaluation for setting reason aside from irrational belief. However, with passing time and accumulated knowledge many accepted concepts turn out to be incorrect or need reevaluation. An example mentioned is the question of earth as the center of our universe. Others come to mind from Euclidean geometry to the nature of the atom.

Euclid's fifth axiom is that through every point it is possible to draw a line parallel to another given line. This eventually turned out to be incorrect when it was realized that space is curved by gravity. The resulting non-Euclidian geometry became of great use and was applied by Einstein in his general theory of relativity. Kant believed that some concepts are a priori and we are born with them: all thoughts would be impossible without them. One of his examples was our intuitive understanding of three-dimensional space based on Euclidean geometry. However, Einstein's space-time fourth dimension superseded Euclidean geometry.

One of the characteristics of intelligent life that developed on our planet is man's unending quest for knowledge. (I am using "man" as a synonym for "humans" without gender differentiation.) When our early ancestors gazed upon the sun and the stars, they were fascinated with these mysterious celestial bodies and their movement. Ever since, man has strived to understand the movement of heavenly bodies. But it was only such pioneers as Copernicus, Kepler, and Galileo who established the concepts of celestial mechanics, which eventually led to Newton's theory of gravitation. Physics thus emerged as a firm science in the seventeenth century.

Contrasted with the mind-boggling scale of the cosmos, our increasing understanding of the atomic nature of matter and the complex world of infinitesimally small subatomic particles and the forces within the atom presents another example for our continuously evolving and therefore changing knowledge. Starting with the early Greek atomists it was believed that the universe was made up of atoms, the further undividable elemental matter. The last century saw, however, an explosive growth in our knowledge of subatomic particles. The recognition of the electron, proton, or neutron was followed by the discovery of quarks and other subatomic particles.

In the nineteenth century, scientists showed that many substances, such as oxygen and carbon, had a smallest recognizable constituent that, following the Greek tradition, they called atoms. The name stuck, although it subsequently became evident that the atom is not indivisible. By 1930, the work of J. J. Thomson, Ernest Rutherford, Niels Bohr, James Chadwick, and others established a solar system-like atomic model consisting of a nucleus containing protons and neutrons and surrounded by orbiting electrons. In the late 1960s, it was further shown that protons and neutrons themselves consist of even smaller particles called quarks. Additional particles in the universe are the electron-neutrino (identical to the electron but 200 times heavier), the muon and an even heavier analog of the electron called tau. Furthermore, each of these particles has an antiparticle identical in mass but of opposite charge. The antiparticle of the electron is the positron (with identical mass but with a charge of +1 instead of −1). Matter and antimatter, when in contact, substantially (but not necessarily completely) annihilate each other. This is the reason why there is extremely little antimatter around and it is so difficult to find it.

Besides particles, the forces of nature play also an increasing role. In the last century, four fundamental forces were recognized: the gravitational, electromagnetic, weak, and strong forces. Of these the weak and strong forces are less familiar, because they are nuclear forces and their strength rapidly diminishes over all but subatomic scales.

During Einstein's time, the weak and strong forces were not yet known. However, gravity and electromagnetism were recognized as distinct forces. Einstein attempted to show that they are really manifestations of a single underlying principle, but his search for the so-called unified field theory failed. So did all efforts till now to combine the two major pillars of modern physics, quantum mechanics, and general relativity. As presently formulated, both cannot be right because they are mutually incompatible. Much effort is going on to find a unified theory for everything, to prove that there is one set of laws for the very large things and the smallest alike, including all forces and particles. Although physicists long believed that the minuscule electrons, quarks, etc., are the smallest particles of matter, the more recently pursued string theory suggests that there is an even deeper structure, that is, each elementary particle is a particular node of vibration of a minute oscillating string. The image replacing Euclid's perfect geometric points is like that of harmoniously thrumming strings (somewhat like Pythagoras' music of the spheres). These infinitesimal loops or strings are suggested as writhing in a hyperspace of 11 dimensions. Of these only four dimensions are easily comprehended by us, the three dimensions of space and Einstein's space-time. The seven additional dimensions of the superstring theory (or as it is sometimes called, the *theory of everything*) are "rolled up" or "compacted" into an infinitesimally small format but are still not dimensionless points. The principle that everything at its most microscopic level consists of a combination of vibrating strands of strings is the essence of the unified theory of all elemental particles and their interactions and thus all the forces of nature.

The complex mathematical basis of the string theory is far beyond the understanding of most of us, and certainly beyond my understanding. However impressive and elegant the mathematical *tour de force* may be that one day probably could produce an "equation for everything" containing 11 dimensions. It is not clear however what its real meaning will be. This is a difficult question to ponder. The tiny domain that superstrings inhabit can be visualized by comparing the size of a proton to the size of the solar system. Our entire solar system is linked on a cosmic scale but to probe the reality of the tiny realm of superstrings would require a particle accelerator 100 light years across. As long the superstring theory or any of its predictions that may emerge cannot be experimentally tested (or disproved), it will remain only a mathematical theory. However, the progress of science may one day result in ingenious new insights that can overcome what we presently perceive as insurmountable barriers.

John von Neuman, one of the greatest mathematicians of the twentieth century, believed that the sciences, in essence, do not try to explain, they hardly even try to interpret; they mainly make models. By a model he meant a mathematical construct that, with the addition of certain verbal interpretations, describes observed phenomena. The justification of such a mathematical construct is solely and precisely that it is expected to work. Stephen Hawking also believes that physical theories are just mathematical models we construct and that it is meaningless to ask whether they correspond to reality, just as it is to ask whether they predict observations.

For a long time, views and concepts (theories) of science were based on facts verified by experiments or observations. A contrary view was raised by the philosopher Karl Popper, according to whom the essential feature of science is that its concepts and theories are not verifiable, only falsifiable. When a concept or theory is contradicted by new observations with which it is incompatible, then it must be discarded. Popper's views were subsequently questioned (Kuhn, Feyerabend) on the basis that falsification itself is subjective; because we do not really know a priori what is true or false. Nonetheless, many consider "scientific proof," that is, verification, still essential. Gell-Mann (Nobel Prize in physics, 1969), for example, writes in his book, *The Quark and the Jaguar*, "sometimes the delay in confirming or disproving theories is so long that their proponents die before the fate of the idea is known. Those of us working in fundamental physics during the last few decades have been fortunate in seeing their theoretical ideas tested during our life. The thrill of knowing that one's prediction has been actually verified and that the underlying new scheme is basically correct may be difficult to convey but is overwhelming." Gell-Mann also wrote "It has often been said that theories, even if contradicted by new evidence, die only when their proponents die." This certainly may be the case when forceful personalities strongly defend their favorite brainchildren. Argumentum ad hominem, however, does not survive for long in science. If a theory is superseded just because its proponent is not around anymore to fend off the others questioning it, it surely sooner or later will be "falsified."

Gell-Mann seems to believe that scientific theories are verifiable and can be proven (confirmed) even in one's own lifetime and thus proven to be true. This is, however, not necessarily the general case. For example, his own quarks turned out not to be the ultimate elementary particles. Recent, experimental observations as well as theory seem to cast doubt on the idea that quarks are indeed the smallest fundamental, indivisible particles of the atoms. They themselves are probably made up of even smaller entities of yet-unknown nature. As discussed, the superstring theory suggests that all matter, including quarks, is composed of vibrating strings. Whereas quarks may stay on for the time being as the fundamental particles, future work probably will bring further understanding of the atomic physics with even more diverse particles and forces being recognized.

"Theory" means the best possible explanation of observations, experimental facts, or concepts (hypotheses) as we know them or conceive them at the time. If new observations (facts or concepts) emerge with which the theory cannot be in accord, then we need to discard or modify the theory. Theories thus cannot be absolutely verified (proven) or even falsified (disproved). This should not imply, however, that a discharged theory was necessarily incorrect at the time it was proposed or represented any intent to mislead or misrepresent. As I have emphasized, science can never be considered without relating it to its historical time frame. There is continuing progress and change in our scientific concepts as new knowledge becomes available. Verification or proof of a theory in the present time thus may be only of temporary significance. Theories can be always superseded by new observations (facts) or concepts. This is the ongoing challenge of science.

The widely invoked concept of "chaos" based on chaotic phenomena is, by our present understanding, unpredictable. According to Ilya Prigogine (Nobel Prize in chemistry, 1977), we have reached the end of certitude in science, which in the future will be increasingly speculative and probabilistic (i.e., ironic). Others, however, feel that eventually a deeper new understanding of some yet-unknown law governing chaotic phenomena will be found. The question is when are we really reaching the limits of real understanding or knowledge? Are vibrating infinitesimally small strings indeed the basis of all matter and forces, allowing a "theory of everything" eventually to be found? Is our universe just one of innumerable multiverses? Is evolution a conscious, predetermined process making the emergence of intelligent beings inevitable or just a consequence of nature? And, ultimately, why is there anything, did it all start and will eventually come to an end or was it always and always will be? Creation means a beginning, but it is possible to think in terms of a continuum without beginning or end. Science in all probability cannot and will never be able to answer these questions. To me, it is only honest to admit that we just do not know.

However, one can go too far in delineating science, as did Thomas Kuhn, who contended that all science reflects not the truth about nature but merely the scientists' prevailing opinion, which is always subject to change. Science has, however, established many fundamental observations and facts of our physical world. For example, atoms exist in a variety corresponding to the elements, so do DNA, bacteria, stars and galaxies, gravity and electromagnetism, natural selection and evolution, etc. Science is our quest for understanding of the physical world, and we should keep this in proper perspective while admitting to the limits of where our human understanding can reach.

A fundamental question in our quest for knowledge and understanding always will be whether there is a higher intelligence beyond our gasp. Many call this "God," but that name invokes very different meanings to different people. It seems that in many ways man created God in his own image or at least depicted him accordingly. Scientists in general find it difficult to believe in something they cannot comprehend or understand. I myself have found it increasingly difficult over the years to believe in supernaturals as proclaimed by many organized religions and their dogmas and regulations. Monotheism is accepted by Judaism, Christianity, and Islam, but there are also other religions practiced by many adhering to Buddhism, Confucianism, Hinduism, Taoism, among others.

The Scriptures of the *Bible*, as well as the *Talmud*, the *Koran*, the *Book of Mormon*, etc., are all valuable teachings and worthy historical documents, but much in them can also hardly be taken verbatim. For example, creation according to the *Book of Genesis* and its limited time line in 6 days cannot be reconciled with scientific knowledge of our physical and biological world. The "days" may indeed mean long periods of time and not our present understanding of days. At the same time, science itself cannot give an answer to how it all started ex nihilo (if it started at all). The "big bang" that happened 12–15 billion years ago only explains how our expanding universe probably started from an immensely dense initial state, not how this came about. We seem to increasingly comprehend how subsequent inflation and continued expansion are governed by physical laws. But there may be innumerable other universes too, which are not necessarily governed by the same physical laws as ours. Universes may not expand forever but eventually collapse (big crunch) into an extremely dense state and eventually gain start for a new big bang. There may be therefore no end or many beginnings.

All recognized religions are by necessity quite contemporary. What is a few thousand years compared to what we know of how long life and even humankind have been around on earth? Disregarding all discrepancy of the biblical act of creation with existing scientific evidence of life on earth, an omnipotent God with a definite act of creation would however greatly simplify many questions for true believers including other questions, such as those of our consciousness and free will, whether there was indeed a beginning, whether there is a reason or goal of our being, and was it planned, to which science itself cannot give answers. I considered myself for long, in Thomas Huxley's terms, as an agnostic. I do not know whether there is a God or creator, or whatever we may call a higher intelligence or being. I do not know whether there is an ultimate cause for our being or whether there is anything beyond material phenomena. I increasingly doubt these things as a scientist, as we cannot prove them scientifically but at the same time we also cannot falsify (disprove) them. For the same reasons, I cannot deny "God" or a higher intelligence with certainty, which would make me a pure atheist. This is not an unusual conclusion reached by many scientists. I simply admit that there is so much that I do not know and that will always remain beyond my (and mankind's) comprehension. I have never had difficulty admitting my limitations (and there are many). Scientists, however, and particularly the more successful ones, are not always prepared to say that there is much they just do not know and that much will stay incomprehensible. In a way, they disregard Kurt Gödel's incompleteness theorem (according to which in mathematics, and thus probably in other sciences, there are insolvable problems) and believe science can eventually provide all the answers. They are consequently tempted to push for justification of their views, their theories, and their assumed proofs. The superstring theory is again an example. One day it indeed may succeed, combining all particles and forces into one complex mathematical equation of 11 dimensions. But what will be its real meaning? Does it necessitate a creator, dealing with an 11-dimensional, highly complex mathematical system in designing the universe? If, on the other hand, there was no higher intelligence and thus no predetermined design, was it as Monod argued only chance that eventually determined the emergence of our universe and our being? Many cosmologists believe in the anthropic principle according to which the universe is as it is because if it were not and only the slightest changes in physical laws had come about, there would be no intelligent life whereby the universe could be known. Knowledge, after all, is our perception of everything. Without existence of intelligent life (as contrasted with lower, primitive forms of living organism) there could not be a quest for understanding of our physical and biological world. The same is true for the cosmos with its innumerable celestial bodies (or innumerable multiverses) and questions of how it all came about and what its purpose and destination are.

In different ways, Monod and Popper suggested that at the interface of reason (consciousness) and the brain a discontinuum must exist. Consciousness (reason) can direct physiological processes in the brain, which in a way denies the principle of conservation of energy because material effects will give further impulses, thus causing either the kinetic or potential energy to increase. The laws of thermodynamics in extreme cases also cannot be valid anymore. This may be the case for the "big bang" conditions of the initial state of a universe, in collapsed stars, or near black holes at enormous densities and pressures, which however can give rise to other new universes in a new entity of planetary system. At the border of mentality, a similar irregulatory (discontinuum) would exist. Thus the physical laws of our universe themselves cannot be considered truly universal. If there indeed are countless other universes (multiverses), their laws could be different and they may remain inaccessible to mankind.

For me, it is not difficult to reconcile science (and by necessity our limited knowledge) and the possibility (although not probability) of a higher intelligence beyond our grasp and understanding. Some call it the reconcilability of science and religion. I would not, however, equate the consideration of a higher being or intelligence with religion. Religion is generally considered as the practice of a belief in a divine power according to specific conduct or rules of mankind. Organized religions

(Christianity, Islam, Judaism, etc.) have difficulties in accepting many scientific facts. For example, the *Book of Genesis* represents a timeline of about 6700 years since the "creation of Man." If we accept fossil evidence and other evidence of evolution (the Pope himself indicated that evolution is indeed probable), this timeline certainly cannot be taken verbatim. Of course the Bible's 6 days of creation may not represent the equivalent of today's earth days but, more probably, periods of possibly very long duration, even billions of years. In any case, evolution cannot, and never attempted to, answer the question of how life originally started. Darwin himself was not initially an atheist but evolution is *per se* not a religion or belief in faith.

Monod's view that it was only chance that brought about life by forming essential building blocks from innumerable individual atoms is in contradiction with mathematical probability. Einstein said that "God is not playing with dice," and himself was a believer. However, it is not necessary to consider that random combination of atoms somehow, despite overwhelming mathematical improbability, resulted in life. We know now that certain relatively simple organic molecules (including amino acids, nucleic acid bases, etc.) could be formed from basic simple inanimate building blocks prevalent in the cosmos and containing only a small number of atoms. It is their combination (and not that of the random combination of all the atoms contained in them) that could have produced our complex biological systems. Spontaneous assembly of some fairly complex molecules using proper templates is now probed by chemists in their laboratories. This, however, would not represent creating life, certainly not intelligent life in a test tube.

When considering how the evolution of life could have come about, the seeding of terrestrial life by extraterrestrial bacterial spores traveling through space (panspermia) deserves mentioning. Much is said about the possibility of some form of life on other planets, including Mars or farther celestial bodies. Is it possible for some remnants of bacterial life, enclosed in a protective coat of rock dust, to have traveled enormous distances, staying dormant at the extremely low temperature of space and even surviving deadly radiation? The spore may be neither alive or completely dead, and even after billions of years it could have an infinitesimal chance to reach a planet where liquid water could restart its life. Is this science fiction or a real possibility? We do not know. Around the turn of the twentieth century Svante Arrhenius (Nobel Prize in chemistry, 1903) developed this theory in more detail. There was much recent excitement about claimed fossil bacterial remains in a Martian meteorite recovered from Antarctica (not since confirmed), but we have no definite proof for extraterrestrial life or its ability to travel through space. In the universe (or multiverse), there indeed may be many celestial bodies capable of maintaining some form of life (probably entirely different from our own terrestrial forms), but will we ever be able to find out about them, not to mention communicate with them? This will remain for a long time only speculation for humankind.

Concerning intelligent life, *Homo sapiens* has been present on earth for only a short period of time of some ten or hundred thousands of years compared to the 4.6 billion-year history of our planet. Simple forms of living organisms were around for billions of years, but man's evolution was slow. At the same time, who can say how long we will be around? According to the theory of Darwinian evolution, species disappear and are replaced by more adept ones. For example, if the dinosaurs had not been made extinct some 65 million years ago when an asteroid hit the earth, mankind probably could not have evolved. The natural selection process continues to go on, and human activities could even accelerate it. Environmentalists argue that all existing species must be preserved, even a tiny fish or a rare bird in remote areas. While we strive to maintain our environment as free from human influence as possible, natural processes inevitably will go on. A major catastrophic event, such as an asteroid colliding with earth, may one day make extinct many of the present life forms. Lower forms of life, such as bacteria, will in all probability survive, and the process of evolution could restart. The resulting higher species, however, may turn out to be different from those that we know today. This may be also the case if intelligent species exist in other parts of the immense cosmos, whose presence, however, we may never be able to ascertain or communicate with. Our limited biological life span, besides other factors, is an obvious limit for deeper space travel and adds to our limitation to "learn everything." We also must realize how fragile and short-lived mankind really is.

When considering the place of science in mankind's overall effort for knowledge and self-expression, it is striking to realize how much inter-relationship exists between our different intellectual activities. Man's drive to express himself can take different forms. Some of these involve making or doing things such as painting, sculpture, architecture, music, writing, drama, dance, and other activities in what we call the arts and letters. The field of humanities in general is concerned with learning about and expressing human thought and relationship. The sciences, as discussed, are concerned with knowledge and understanding of nature, the physical world, and the forces, laws, and rules governing them. For science to move forward to new levels of understanding, we need to advance creative new ideas, concepts, and theories and to explore their reality. In this sense, science essentially is not much different from the concepts or thoughts of the humanities and letters or forms of self-expression in the arts, music, etc. Arts, humanities, and creative sciences are closely related, even if this is not always fully recognized. Of course, we must differentiate the artist from the artisan, the composer or cre-ative musician from the mere practicing performer, as much as the "reg-ular" scientist or technologist from the creative one ("revolutionary" in Thomas Kuhn's sense).

I was teaching in the past a freshman seminar on the relationship of the sciences with the humanities and economics. One semester of discussions led together with an economist and a humanist colleague (Morty Shapiro and Sally Pratt) was centered on Goethe's Faust as an example of the interrelationship between our seemingly unrelated fields. Faust is an epitome of Goethe's life experience. He was a great poet, but also a remarkable man with wide interest in different fields including the sciences and economics (he was for a while the finance minister of his German principality). The first part of the Faust story can be looked at as the story of an alchemist who strives through the philosopher's stone to make gold. Even as he fails, in the second part of the Faust story Goethe discusses paper money as a way achieving this goal. Paper money assumes the role of gold and even creates new capital and wealth. Goethe thus showed the interrelationship of economics with alchemy. The story of Faust also gives us a sense of the state of alchemy (i.e., early chemistry) in that period of history.

There are many other examples of interrelationship. "Symmetry," for example, is of fundamental importance in the sciences and arts alike. It plays a key role in our understanding of the atomic world as well as the cosmos. The handedness of molecules with nature selecting one form over the other contributes fundamentally to the evolution of living organisms. Symmetry also plays a significant role in typographical number theory in mathematics. It is also of great significance in the complex string theory through which mathematical physics is trying to develop a "theory for everything." At the same time there is a key role of symmetry in the arts expressed by varying examples from the fascinating paintings and graphics of Escher to Bach, who in some of his sonatas managed two (or more) musical lines going simultaneously, in a way creating a musical symmetry effect.

As science in some way or other affects practically all aspects of our life, without necessarily attempting to give a deeper understanding of its complexities, it is essential that all educated people in the modern world should have at least a rudimentary education in science. Literacy should not only mean being able to read and write (or use a computer) but also having at least a minimal "science literacy." At all levels of science education, the clarity of presenting facts and concepts is of great importance but should not be at the expense of accuracy. This is not easy, because science should be also presented as a fascinating, dynamic, and challenging topic that should catch the attention of children and adults alike and inspire them to follow up with more detailed studies and reading. This perhaps explains my lifelong devotion to science and teaching alike.

The twentieth century was considered the century of science and technology. It produced many renowned scientists, some of whom, such as Albert Einstein, gained wide general recognition. Science education, at the same time in the post-Sputnik second half of the century, started to lose some of its shine and cultural significance. It is difficult, however, to imagine how tall edifices can be built without proper foundations. The increasing interest in interdisciplinary studies also puts frequently premature emphasis on crossing different disciplines without first establishing solid foundations in them. Many learn readily the vocabulary and superficial aspects of a field but lack solid grounding and knowledge. There is the danger that the underpinnings essential for science will be weakened. The trailblazers of the DNA revolution, Crick and Watson, are household names of twentieth century science. At the same time, nature does not readily or frequently gives such intuition and recognition. It is necessary to be well prepared and able to pursue the science of frequently broad and complex fields with all the skills and knowledge acquired through solid education. In my field of chemistry, "magic" certainly comes only very rarely and even then generally only coupled with consistent, hard work and study. Ideas, of course, are the essence of new discoveries, but at the same time one must be well prepared to realize which has merit and significance, as well as be able to stay the course to follow them through. Despite general assumption, there is usually little glamour in science compared with the long and frequently disappointing efforts it demands. There is, however, the occasional epiphany of discovery and fundamental new understanding, the eureka or ecstasy that makes it all worthwhile, which only those who have experienced it can really appreciate.

· 3 ·

Chemistry:
The Multifaceted Central Science

As a chemist, I should briefly discuss what my field of science is. Here, I also reflect on its historical development and scope, which help to put in perspective the broad background on which our contemporary chemistry was built, and where my own work fits in.

Chemistry deals with substances, their formation, and subsequent transformations as well as their composition, structure, and properties. Chemistry does not deal with either the infinitely small world of subatomic physics or the cosmological mysteries of the infinitely large cosmos, although extraterrestrial chemistry is involved in the material universe. Chemistry does not directly deal with the living world, but it is essential for our continued understanding of the world at the molecular level. Chemistry is thus also essential to our understanding of other sciences and is recognized to be the central science bridging physics and biology, drawing on the basic principles of physics while enabling us to understand biological systems and processes at the molecular level.

The concepts of chemistry were formulated on the observation and study of various elements and their compounds. Matter was suggested to be composed of indivisible particles called atoms by the ancient Greek philosophers. In modern times, more than 100 different kinds of atoms are recognized, composing the chemical elements. When atoms combine, they form molecules and compounds (an assembly of a large number of molecules). They are held together by forces generally referred to

A Life of Magic Chemistry: Autobiographical Reflections Including Post-Nobel Prize Years and the Methanol Economy, Second Updated Edition. George A. Olah with Thomas Mathew.
© 2015 John Wiley & Sons, Inc. Published 2015 by John Wiley & Sons, Inc.

as chemical bonding. In the strict sense, no such thing as the "chemical bond" exists, only atoms held together by sharing electrons in some way. The highest probability of the location of electrons between atoms is depicted by the chemist by two-, three-, or multicenter bonding (sharing electrons). The transformation of molecules and compounds by various changes leads to new and different molecules.

Because of chemistry's very wide scope, it is customary to divide it into branches. One of the main branches is *organic chemistry*, which originally dealt with compounds that were obtained from (or related to) living organisms but is now generally recognized to be the chemistry of the compounds of carbon or, more precisely, of hydrocarbons (compounds of carbon and hydrogen) and their derivatives. *Inorganic chemistry* deals with compounds of the elements other than organic compounds (i.e., hydrocarbons and their derivatives). In *biochemistry*, the compounds and chemical reactions involved in processes of living systems are studied. *Biological chemistry* (more recently also *chemical biology*) involves the chemistry of biological systems. *Physical chemistry* deals with the structure of compounds and materials as well as the energetics and dynamics of chemical changes and reactions. It also includes related theoretical studies (*theoretical chemistry*). *Analytical chemistry* encompasses the identification and characterization of chemical substances as well as their separation (isolation) from mixtures.

There is an ever-increasing number of further subdivisions, or what I would call "hyphenated" branches of chemistry or chemically related sciences. Whether "chemical-physics" or "chemical-biology" is more meaningful than "physical chemistry" or "biological chemistry" may depend on the point of view one wants to look from.

Chemistry is the science of molecules and materials, physics deals with forces, energy, and matter (also including fundamental questions of their origin), and biology deals with living systems. Chemistry deals with how atoms (formed from the original energy of the big bang) build up molecules and compounds, which eventually organized themselves into more complex systems of the physical and biological world. It also deals with man-made compounds and materials. Chemistry is not directly concerned with such fundamental questions as how the universe was formed, what (if any) the origin of the big bang was, what the nature of the infinite miniscule subatomic world is, or, on the other hand, the dimensionless cosmos, how intelligent life evolved, etc. It deals with molecules composed of atoms of the elements and their assembly into materials or biological systems with their eventual enormous complexity. It is frequently said that in the unofficial order of the sciences physics comes first (mathematics is not considered strictly a science but rather a way of expression of human knowledge, not unlike the languages), followed by chemistry and then biology. This is also, incidentally, the sequence in which the prizes are presented during the Nobel ceremony, although there is no prize for biology as such, only medicine or physiology (more about this in Chapter 11).

What we now call chemistry slowly emerged over the centuries as mankind's use of varied substances and compounds and the quest for understanding of the material world evolved. The practical beginning of chemistry goes back to ancient Egypt, based on experience gained in metals, glass, pottery, tanning, and dying substances, etc. On the other hand, speculations by the Greeks and peoples in the East laid the foundation of this quest of better understanding into the nature of the material world.

It was in the great school of Alexandria that these separate paths came together and eventually led to the alchemists and iatrochemists of future generations and, eventually, the chemistry of modern science.

In all of the early natural philosophies, there is the underlying idea that there was some primordial element or principle from which the universe was derived. It was perhaps Thales who in his doctrine first speculated that water was the prime element. Plato, in his Timotheus based on Aristotle, suggested that four elements made up all things in the universe: earth, water, air, and fire. These platonic elements were assigned characteristic geometric shapes. The elements were mutually transformable by breaking down their geometric shapes into those of the others. The doctrine of the four elements was further taught by Aristotle, who emphasized the broad principle that one kind of matter can be changed into another kind; that is, transmutation is possible. Aristotle's concept differed fundamentally from that of the unchangeable elements (Empedocles) and the mechanical hypothesis of Democritus, according to which the world was built upon the meeting of rapidly moving atoms, which themselves, however, are of unalterable nature. In Egypt and the area of Mesopotamia where working of metals was advanced, these concepts gained roots. When the Arabs conquered Egypt in the seventh century and overran Syria and Persia, they brought a new spirit of inquiry onto the old civilization they subdued.

To the development of what eventually emerged as the science of chemistry, metallurgy and medicine made contributions as well, but these origins have not crystallized into a unified picture. The fourth major contributor was alchemy, which originated in Egypt and the Middle East and had a twofold aspect. One aspect was practical, aimed at making gold from common base metals or mercury and thus providing unlimited wealth for those who could achieve it. The other aspect was the search in the medieval world for a deeper meaning between man and the universe.

These days, we consider alchemy a strange and mystical mixture of magic and religion, at best an embryonic form of chemistry but more a pseudo-science. But as Jung pointed out, alchemy was not simply a futile quest to transform base metals into noble gold. It was an effort in a way to "purify the ignoble and imperfect human soul and raise it to its highest and noblest state." It was thus in a way a religious quest–not necessarily just a scientific one. Matter and spirit were inseparable to medieval alchemists, and they strove to transform them through these procedures, which amounted to sacramental rites and religious rituals as much as scientific research.

This is not the place to discuss the frequently reviewed historical and philosophical aspects of alchemy, but it is worthwhile to recall some rather late adherence to the precepts of alchemy by giants of the human intellectual endeavor. Johann Wolfgang Goethe is best known for his poetry and literature as the author of Faust. He himself, however, considered some of his major achievements to be in science. His interests were varied but mainly related to chemistry. He developed an early interest in alchemy, which, however, he overcame in later life. Goethe's classic character Faust reflects his fascination with the alchemist's effort to produce gold but eventually recognizes its futility and failure.

Newton, one of the greatest physicists of all time, is said to have spent more time on his alchemist efforts and experimentation than on his physical studies. Was this only scientific fascination, or had his position as Master of the Royal Mint added to his interest? Not only could he have wanted to safeguard the purity of the gold coins of the realm, but the thought may have crossed his mind that perhaps he could also produce the gold itself by transforming much less valuable mercury. It is, however, also possible that he was looking for an understanding of the elemental matter and possible transformation of elements. Newton also failed in his quest, and it was only in our nuclear age that transmutation of the elements was achieved. In fact, in 1980 a bismuth sample was transmuted into a tiny amount of gold in the Lawrence-Berkeley Laboratory, although at a very high cost.

In the narrow sense of the word, alchemy is the pretended art of transmuting base metals such as mercury into the noble ones (gold and silver). Its realization was the goal up to the time of Paracelsus and even later. Alchemy in its wider meaning, however, stands for the chemistry of the middle ages. Alchemy thus in a sense focused and unified varied and diverse chemical efforts, which until that time were disconnected and focused on producing varied practical materials for human needs. Alchemy indeed can be considered an early phase of the development of systematic chemistry. As Liebig said, alchemy was "never at any time anything different from chemistry."

It also must be noted that the processes described by the alchemists going back to the thirteenth century were not considered to be miraculous or supernatural. They believed that the transmutation of base metals into gold could be achieved by their "art" in the laboratory. But even among the late Arabian alchemists, it was doubted whether the resources of the art were adequate to the task. In the West, Vincent of Beauvais already remarked that success had not been achieved in making artificial metals identical with the natural ones. Roger Bacon, however, still claimed that with a certain amount of the "philosopher's stone" he could transmute a million times as much base metal into gold.

In the earlier part of the sixteenth century, Paracelsus gave a new direction to alchemy by declaring that its true object was not the making of gold but the preparation of medicines. This union of chemistry with medicine was one characteristic goal of iatrochemists, of whom he was the predecessor. The search for the elixir of life had usually gone hand in hand with the quest for the philosopher's stone. Increasing attention was paid to the investigation of the properties of substances and of their effects on the human body. Evolving chemistry profited by the fact that it attracted men who possessed the highest scientific knowledge of the time. Still, their belief in the possibility of transmutation remained until the time of Robert Boyle.

It was indeed in the seventeenth century that chemistry slowly emerged in its own right as a science. Robert Boyle probably more than anybody else paved the way helping to disperse its reputation as a tainted alchemical pseudo-science. In his book, *The Sceptical Chymist*, he emphasized the need to obtain a substantial body of experimental observation and stressed the importance for the quantitative study of chemical changes. Boyle is remembered for establishing that the volume of gas is inversely proportional to its pressure and for his pioneering experiments on combustion and calcination. He pointed out the importance of working with pure, homogeneous substances, and thus in a way he formulated the definition, but not necessarily the concept, of chemical elements, which he believed to be "primitive and simple, or perfectly unmingled bodies, not being made of any other bodies." However, he still believed that water, air, and fire were elementary substances. Nonetheless, Boyle believed in the atomic theory and that chemical combination occurs between the elementary particles. He also had good ideas about chemical affinity. His followers (Hooke, Mayow, et al.) extended his work.

Boyle and his followers represented the English school of chemistry, which, however, declined by the end of the seventeenth century, leading to the revival of the German iatrochemical school and introduction of the phlogiston theory. Iatrochemists believed that chemical substances contained three essential substances: sulfur (the principle of inflammability), mercury (the principle of fluidity and volatility), and salt (the principle of inertness and fixity). Becker, around the end of the seventeenth century, modified these three general constituents to represent terra lapida corresponding to fixed earth present in all solids (i.e., the salt constituent of the iatrochemists), terra pinguis, an oily earth present in all combustible materials (i.e., the sulfur constituent), and terra mercurialis, the fluid earth corresponding to the mercury constituent. Stahl subsequently renamed the terra pinguis "phlogiston," the "motion of fire" (or heat), the essential element of all combustible materials. Thus the phlogiston theory was born to explain all combustion and was widely accepted for most of the eighteenth century by, among others, such luminaries of chemistry as Joseph Priestley.

Isolation of gases from calcination of certain minerals (fixed air) and from the air itself represented the next great advance in chemistry. Cavendish studied the preparation of hydrogen, "inflammable air," as he termed it. Priestley in the 1770s discovered and isolated several gases, namely, ammonia, hydrochloric acid gas, various nitrogen oxides, carbon monoxide, sulfur dioxide, and, most notably, oxygen, which he considered "dephlogisticated air." Independently, and even somewhat before Priestly, the Swedish apothecary Scheele discovered oxygen and pointed out that air could not be an elementary substance as it was composed of two gases, "fire air" or oxygen and "foul air" or nitrogen, according to a ratio of one to three parts by his estimate. However, Scheele still believed in the phlogiston theory, and he thought that oxygen's role was only to take up the phlogiston given out by burning substances. It was Antoine Lavoisier in France working in the latter part of the eighteenth century along rather different lines who systematically criticized the prevailing traditional chemical theories of his time. He realized that Priestley's "dephlogisticated air" was the active constituent of air in which candles burned and animals lived. Metals absorbed it on calcination. In 1775 Lavoisier thought that oxygen was the pure element of air itself, free from "impurities" that normally contaminated it. Scheele, however, showed in 1777 that air consisted of two gases, oxygen (which supported combustion) and nitrogen (which was inert). Lavoisier accepted Scheele's view and suggested in the next year that the atmosphere was composed of one-quarter oxygen and three-quarters nitrogen, a ratio that was subsequently corrected by Priestley to one-fifth oxygen and four-fifths nitrogen. Clearly, the discovery of oxygen and its role in combustion played a most significant role in the development of chemistry.

In 1783, Lavoisier announced a basic reevaluation of the "chemical theory," rejecting completely the phlogiston theory. At the same time, he elevated oxygen to a general explanatory principle (in a manner reminiscent of the iatrochemists), ascribing to it properties that were not experimentally warranted. For example, he suggested that oxygen was the basis of the acidifying principle, all acids being composed of oxygen united to a nonmetallic substance. This was subsequently disproved by Humphry Davy, who in 1810 showed that hydrochloric acid did not contain oxygen. Lavoisier himself followed up his studies on combustion, discrediting the phlogiston theory with an effort to put chemistry on firm ground by suggesting an entirely new nomenclature to bring the definitions derived from increasing experimental facts into the general context of chemistry. The *Méthode de Nomenclature Chimique*, published in 1787, introduced names for 33 "simple substances" (i.e., elements) including oxygen, nitrogen (azote), hydrogen, etc., named acids, and their derived substances, that is, salts. Through this new naming system (much of it still in use) Lavoisier also put the principles of chemistry on which it was based into common practice. He followed up with his famous *Traité Élementaire de Chimie*, a book that broke with traditional treatises. He discussed chemistry based on ideas backed by facts proceeding according to "natural logic" from the simple to the complex. He presented chemistry on the basis of analytical logic. He also broke with traditional historical pedagogy, according to which the three realms of nature were: mineral, vegetable, and animal. Lavoisier's treatise was the first modern work of chemistry and his major achievement in the "chemical revolution" he started. It is an irony of fate that this revolutionary chemist some years later lost his life to the guillotine of the French Revolution.

The dawn of the nineteenth century saw a drastic shift from the dominance of French chemistry to first English-, and, later, German-influenced chemistry. Lavoisier's dualistic views of chemical composition and his explanation of combustion and acidity were landmarks but hardly made chemistry an exact science. Chemistry remained in the nineteenth century basically qualitative in its nature. Despite the Newtonian dream of quantifying the forces of attraction between chemical substances and compiling a table of chemical affinity, no quantitative generalization emerged. It was Dalton's chemical atomic theory and the laws of chemical combination explained by it, which made chemistry an exact science.

In 1804, Dalton formulated his concept that identified chemical elements with atoms. The notion of atoms, the smallest corpuscles of matter, was not new, of course, and had been around in some form or other since antiquity. Dalton, however, addressed the question to differentiate atoms not only by their size (or shape) but also by their weight. To do this, Dalton turned to Proust's law, according to which the relationship of masses of two or more elements, which combine is fixed and not susceptible to continuous variation, and made it the center of his atomic hypothesis. He suggested that chemical combination takes place via discrete units, atom by atom, and that atoms of each element are identical. He also added the concept of multiple properties, that is, when two elements form different compounds the weights in which one element will combine with another are in a simple numerical ratio. Dalton's atomic concept gave the whole body of available chemical information an immediate, easily recognizable meaning. What was also needed, however, was to relate all the atomic weights to a single unit. Dalton chose the atomic weight of hydrogen for this unit. Dalton's atoms also differed fundamentally from Newtonian corpuscles because they were not derived from an attempt to be based on the laws of motion and the attraction of single bodies whose ultimate constituents would be atoms.

Shortly after publication of Dalton's *New System of Chemical Philosophy* Gay-Lussac announced his observations that "volumes of gas which combine with each other and the volume of the combination thus formed are in direct proportion to the sum of the volumes of the constituent gases." The volumetric proportions of Gay-Lussac and Dalton's gravimetric ratios indeed supplement each other, although they themselves mutually contested and rejected each other's concepts.

Whereas most chemists focused their attention on speculation about atoms and the question of atomic weights, the constant multiplicity in compounds occupied an increasingly central role. The new concept of substitution, that is, the replacement of one element by another in a compound, started to make a major impact on chemistry in the 1840s. It was probably Dumas, who in the 1830s at the request of his father-in-law (who was the director of the famous Royal Sèvres porcelain factory) resolved an event that upset a royal dinner party at the Tuilleries in Paris. During the dinner, the candles in the hall started to emit irritating vapors, forcing the king and his guests into the gardens. Dumas found that the vapors were hydrochloric acid, which resulted from the replacement of hydrogen atoms by chlorine atoms during the bleaching of the wax of which the candles were made and subsequent decomposition. Dumas related this to other previous observations (by Faraday, Gay-Lussac, Liebig, and Wöhler) and his own study of the chlorination of acetic acid, establishing the principle of substitution.

The development of chemistry in the rest of the nineteenth century saw the emergence in Germany of such dominant persons as Wöhler and Liebig, who, helped by the work of the Swedish chemist Berzelius, spearheaded the emergence of organic chemistry. In 1828, Wöhler prepared urea by the rearrangement of ammonium cyanate (an inorganic salt) in aqueous solution upon heating. Up to this time, "organic" compounds had been obtained only from living organisms. Wöhler and Liebig's work preparing a whole series of these compounds by combining organic and inorganic components fundamentally changed this situation. Organic chemistry, that is, the chemistry of carbon compounds, emerged and gained ever-increasing significance.

Liebig's students followed and greatly extended this new trend. My purpose here is not to trace the history of organic chemistry in any comprehensive way. Giants such as Bunsen, Kolbe, Baeyer, and Emil Fischer built a powerful tradition of synthetic and structural organic chemistry in the nineteenth century but at the same time tended to show significant antipathy toward any particular theory, despite the fact that physical chemistry created at the same time by Oswald and others made great progress.

Liebig's student, Kekulé, however, was the most significant exception. His systematic classification of organic compounds led him to the realization of carbon's attribute of combining with four atoms or groups, just as oxygen binds to two atoms or groups. However, it probably was Frankland who first recognized the "saturation capacities of elementary atoms and the abilities of polyvalent atoms to couple in characteristic ways." Couper as well as Butlerov did much to point the way to express logical structural formulas. It was, however, Kekulé who in the 1860s developed more fully the concept of valence based substantially on his realization of the valence of four of carbon in its compounds. It was a century later that I was able to show that carbon in some systems can also bind simultaneously five, six, or even seven atoms or groups, introducing the concept of hypercarbon chemistry (see Chapter 10.)

With the continued discovery of new elements in the nineteenth century, chemists started to group the elements together according to some empirical orders, which eventually developed into the Mendeleev periodical table. Mendeleev published his text, *Principles of Chemistry* around 1870. Previously, Gerhardt had ordered most abounded chemistry around three typical molecules, H_2, H_2O, and NH_3, to which Kekulé added a fourth, CH_4. These molecules were seen to show a scale of equivalent valence ranging from one to four (i.e., H^IH, $O^{II}H_2$, $N^{III}H_3$, and $C^{IV}H_4$). The "typical" elements contained in this classification and their valences were well suited to a logical order of presentation. Because Gerhardt also included HCl as a typical molecule and this acid readily formed salts with alkali metals, Mendeleev in his first volume of *Principles*, treated H, O, N, and C together with the four known halogens and alkali metals. Mendeleev's typical elements were all of low atomic weight and widely distributed in nature, thus being of organic types and not representing all elements. In his second volume, he added metals and began to order the elements according to their valence and atomic weight. The horizontal relationships related to valence, with a broad "transition" group following the alkali metals. Mendeleev was able to slowly fill gaps in his system, whereas other predicted "vacancies" were subsequently filled by discovery of new elements.

For nearly half a century, Mendeleev's periodical table remained an empirical compilation of the relationship of the elements. Only after the first atomic model was developed by the physicists of the early twentieth century, which took form in Bohr's model, was it possible to reconcile the involved general concepts with the specificity of the chemical elements. Bohr indeed expanded Rutherford's model of the atom, which tried to connect the chemical specificity of the elements grouped in Mendeleev's table with the behavior of electrons spinning around the nucleus. Bohr hit upon the idea that Mendeleev's periodicity can be explained by the limited number of electrons occupying the same orbital. When the orbital is filled, one moves down a line in the table.

Chemists were quick to appreciate Bohr's model because it provided an extremely clear and simple interpretation of chemistry. It explained the reason behind Mendeleev's table, that is, the position of each element in the table is nothing other than the number of electrons in the atom of the element, which, of course, represents an equal number of periodic changes in the nucleus. Each subsequent atom has one more electron, and the periodic valence changes reflect the successive filling of the orbital. Bohr's model also provided a simple basis for the electronic theory of valence.

Physicists traditionally pay little attention to the efforts of chemists. One can only wonder whether, if physicists had given more consideration to the chemists' periodical table, they could have advanced their concept of atomic theory by half a century! This is the irony of divided science, and it shows that chemistry is not a simple derivative of physics. It should be emphasized that all of science represents one entity. The understanding of the laws and principles of physics provides the foundation of our understanding not only of the limitless universe but also of the small elemental particles, the atoms that are the building blocks of the molecules and compounds of the chemist. These laws and principles also explain the forces that combine atoms (referred by chemists as chemical bonds) as well as the physical basis of valence. After Dirac derived his relativistic quantum mechanical equation for the electron as well as the physical basis of valence in 1928, he was said to have remarked that these, together with Schrödinger's equation describing H_2, explained all of chemistry, which can be simply derived from the first principles of physics. What he did not emphasize was that similar treatment from first principles of more complex, large molecules and those of nature alike would probably remain a challenge for a very long time.

For two thousand years, the atoms were considered the smallest and indivisible units of nature. At the beginning of the nineteenth century, Dalton got chemistry on the path of atomic theory with his book, *A New System of Chemical Philosophy*, in which he argued that unbreakable atoms form compounds by linking with other atoms in simple whole-number proportions. It was at the end of the nineteenth century that Thomson discovered the existence of particles a thousand times smaller than the smallest atom. When it was found that these particles carry negative charge and are the fundamental unit of electricity, they were called electrons. Because, however, under normal circumstances atoms have no overall charge, there had to be a positive electrical component to neutralize the electrons. Rutherford subsequently showed that the center of the atom is a very dense nucleus that carries a positive charge. The negative electrons are orbiting, at substantial distance, the otherwise empty space of the atom. Rutherford's dynamic orbiting model, however, was unable to answer the question of why the moving charged particles (electrons), according to classical requirements of physics, do not lose energy and collapse into the nucleus. It was Bohr who subsequently linked Rutherford's atom with Planck's concept that energy, not unlike heat or light, is not continuous as Newton thought but exists in discrete quanta. Bohr developed his model of the atom in which electrons travel around the nucleus in circular orbits but only certain-sized orbits are possible, and these are determined by quantum rules.

The development of the structural theory of the atom was the result of advances made by physics. In the 1920s, the physical chemist Langmuir (Nobel Prize in chemistry, 1932) wrote, "The problem of the structure of atoms has been attacked mainly by physicists who have given little consideration to the chemical properties which must be explained by a theory of atomic structure. The vast store of knowledge of chemical properties and relationship, such as summarized by the Periodic Table, should serve as a better foundation for a theory of atomic structure than the relativity meager experimental data along purely physical lines."

At the heart of chemistry, as Langmuir pointed out, was the periodic table. As mentioned since 1860, chemists had realized that although the elements could be arranged according to their increasing atomic weights, they contained properties that showed periodic similarities. The periods seemed to be eight elements long and then to repeat. The number eight seemed to be crucial, but no explanation was given until 1916, when G. N. Lewis explained this by his "octet rule" based on a cube-like model of the atoms, the electrons being equidistant from the nucleus. Moving up through the periodic system, new cubes were simply added.

Langmuir further developed the pioneering ideas of G. N. Lewis to bring chemistry in line with the new physics. Their combined concept also introduced the shared electron pair as the basis of chemical bonding. Sharing electrons was the glue, they said, that held molecules together. (It is regrettable that Lewis, unlike Langmuir, did not receive the Nobel Prize in recognition of his fundamental contributions).

Lewis–Langmuir covalent electron pair bonding, combining two atoms by sharing two electrons (2e–2c bond) became a foundation of our understanding of chemical bonding. Two electrons, however, can be shared simultaneously not only by two atoms but, as shown in Lipscomb's extensive studies of boron compounds (Nobel Prize in chemistry, 1976), also by three atoms (2e–3c bond). It was my good fortune to realize that similar bonding is also possible in general with electron-deficient carbon (see Chapter 10). Kekulé's original concept that carbon can bind no more than four other atoms was thus extended into the new area of hypercarbon chemistry.

Chemistry as it was realized substantially derives from the interaction of electrons. The electronic theory of chemistry, particularly of organic chemistry, emerged, explaining the great richness of chemical observations and transformation, as expressed by Ingold, Robinson (Nobel Prize in chemistry, 1947), Hammett, and many others following in their footsteps.

The advent of nuclear chemistry of the twentieth century not only allowed the creation of scores of new radioactive elements but also gave an explanation for how the formation of hydrogen and helium after the initial big bang was followed by the formation of the other elements. In accordance with Einstein's realization, energy equals matter (although terrestrial experimental verification was only recently obtained at the Stanford accelerator when it was shown that particles indeed are formed from high energy). According to the "big bang" theory, all the matter in the universe initially was contained in a primeval nucleus of enormous density and temperature, which then exploded and distributed matter uniformly through space. As the temperature fell from an initial estimate of 10^{32} K after 1 second to 10^{10} K, elemental particles (neutrons, protons, and electrons) formed. Upon further cooling, conditions became right to combine these particles into nuclei of hydrogen and helium: the process of element building began. From 10 to 500 seconds after the big bang, the whole universe is thought to have behaved as an enormous nuclear fusion reactor. The composition of all atoms in the universe is suggested to be about 88.6% hydrogen, whereas helium makes up 11.3%. Together they account for 99.9% of the atoms.

When hydrogen is burned up in the nuclear furnace of a star, helium burning takes over, forming carbon, which in turn leads to oxygen, etc. Subsequent emission processes releasing α-particles, equilibrium processes; neutron absorption, proton capture, etc., lead to heavier elements.

Chemists are satisfied how atoms of the different elements could form from the initial enormous energy of the big bang explosion, without, however, the need to concerning themselves with the reason for its origin. Atoms subsequently can combine into molecules, which in turn build increasingly complex systems and materials, including those of the living systems. This is the area of the chemists' interest.

The second great advance of twentieth century physics came in the form of quantum theory. Quantum theory deals only with probabilities. The great success of quantum mechanics and its wide applications cannot overcome this point. It was because of this that Einstein himself never fully accepted quantum mechanics (his frequently quoted saying was "God does not play dice"). He attempted for many years to develop a theory combining quantum mechanics and relativity but never succeeded. The effort still goes on for a unified theory, and it may one day succeed. As mentioned, the string theory promises to develop a complex mathematical solution with some 11 dimensions, but it is unclear what new physical meaning will derive from it. For chemists, the advent of quantum theory also brought new vistas. It was Pauling and subsequently others who introduced into chemistry the concepts of quantum mechanics. Pauling, for example, treated the structure of benzene in terms of what he called "resonance" (instead of "electron exchange" used by Heisenberg, Heither, London, and others). He assumed limiting forms to derive the more stable intermediate structure of higher probability, in a way mixing assumed single and double bonds in benzene, leading to stabilized partial double bonds with their peculiar properties. Pauling's resonance theory was much debated. In the 1950s, it was strongly criticized in a pamphlet by Soviet scientists, with the approval of Stalin himself, which claimed that it was antimaterialist and anti-Marxist, because the limiting valance bond resonance "structures" had no physical reality. These critics, however, showed no understanding for the probability nature of quantum theory. (Some Russian scientists such as Syrkin and Dyatkina fully understood it, but were ostracized at the time for it.) When molecular orbital concepts were more widely introduced, the "controversy" faded away. Increasingly, the use of quantum mechanical calculational methods, first based only on approximations and later on more precise ab initio methods (Pople, Nobel Prize in 1998), allowed chemists (Schleyer and many others) to use calculational chemistry as a powerful tool to theoretically calculate not only energies but structures and expected reaction paths. Experimental chemistry, however, still remains the essence of chemistry, but computational methods greatly supplement it and in some cases point the way to new understanding and even unexpected chemistry. Organic chemistry in general made great strides. Synthetic methodology (Barton, Nobel Prize in 1969), including the stepwise preparation of complex molecules (frequently those of natural products or biologically relevant systems), made and continues to make spectacular advances. Woodward's (Nobel Prize in 1965) mastery of such syntheses is legendary, as is Corey's (Nobel Prize in 1990) retrosynthetic approach, building molecules by reassembling them from their derived simpler building blocks. Asymmetric synthesis (i.e., that of chiral, optically active molecules) also advanced dramatically (Kagan, Nayori, Sharpless, and others). So did the synthesis and study of an ever-increasing array of inorganic compounds, including main group elements, organometallics (Fischer, Wilkinson, Nobel Prize in 1973), and metal–metal bound systems (Cotton).

Chemistry also contributed in a major way to the development of modern biological sciences through an ever more sophisticated understanding at the molecular level. Long are gone the days when Emil Fischer, who can be credited as having established biochemistry through his pioneering chemical work on carbohydrates, amino acids, peptides, and his "lock and key" concept of enzyme action, practically gave it away, saying that proper chemists should not deal with compounds they cannot properly isolate, purify, crystallize, and analyze by the classical methods of his time. Chemistry has provided methods to isolate even the most complex bio-organic molecules, using a whole array of efficient separation and analytical (spectroscopic) methods, and in many cases even synthesize them, paving the way to molecular biology and biotechnology. X-ray crystallography of isolated and crystallized DNA (by Rosalind Franklin) made it possible for Crick and Watson (Nobel Prize in medicine, 1962) to realize the double-helix structure of DNA (in contrast to Pauling's intuition of a triple helix, which in this case proved to be wrong).

I do not wish to go into further discussion of the only too well-known close interrelationship of chemistry and biology, which some these days like to call chemical biology instead of biological chemistry. The interface of chemistry and physics can be equally well called chemical physics or physical chemistry, depending on from which side one approaches the field. It is important to realize that chemistry occupies a central role between physics and biology. Chemistry is a truly central, multifaceted science impacting in a fundamental way on other sciences, deriving as much as it contributes to them.

· 4 ·

Growing Up in Hungary and Turning to Chemistry

I was born in Budapest, Hungary on May 22, 1927. My father, Gyula Olah, was a lawyer. My mother, Magda Krasznai, who came from a family in the southern part of the country, when it became part of Yugoslavia fled to the capital, Budapest, at the end of World War I.

Growing up in Budapest between the two World Wars in a middle-class family provided a rather pleasant childhood. Budapest was and still is a beautiful city. It is the only major city on the Danube through the middle of which the river flows in its true majesty. You can be in Vienna without necessarily realizing that it is also located on the Danube, as the river flows only through the outskirts. Not in Budapest! The half-mile-wide river divides Budapest into the hilly side (Buda) and the flat city (Pest). In fact, Pest and Buda were separate cities until 1873, when a permanent bridge was built to combine them. It is a beautiful chain bridge, and it still stands, after having been rebuilt following its destruction, together with all other bridges, at the end of World War II.

I was born in my parents' apartment in the Pest part of the city off the famous Andrassy Avenue and across from the Opera House. The apartment building at 15 Hajos Street was (and still is) an impressive building, designed by Miklós Ybl, a noted architect who designed the Opera House and other major public buildings. Many members of the Opera Company, from conductors to musicians to singers, were tenants in the same building. The composer Miklós Rózsa, who initially

A Life of Magic Chemistry: Autobiographical Reflections Including Post-Nobel Prize Years and the Methanol Economy, Second Updated Edition. George A. Olah with Thomas Mathew.
© 2015 John Wiley & Sons, Inc. Published 2015 by John Wiley & Sons, Inc.

became famous for his musical scores for Alexander Korda's movies, grew up there, brought up by his uncle who was an orchestra member. I met him many years later in Los Angeles, and we reminisced about it.

Quite naturally, I attended many opera performances and rehearsals with neighbors and friends from early childhood on. I still remember a rehearsal conducted by a gentle, smallish man who became, however, a holy terror with the orchestra and the singers alike once he raised his baton. His name was Arturo Toscanini.

I myself, however, had no musical inclination or talent. It was consequently somewhat amusing that years later, at the end of the war in the spring of 1945, friends made me a "Member of Budapest Opera House," allowing me to carry a rather official-looking identification card which was even respected by the occupying Soviet military and entitled me to some privileges. (Many civilians of all ages, including survivors of the Nazi terror, were taken indiscriminately into captivity as so-called "prisoners of war," frequently never to be heard from again.) My duties during my short-lived "operatic" career, however, included only such chores as clearing rubble or moving around pianos and other heavy musical instruments.

With my parents and brother in 1928

At the end of World War I, Hungary lost more than half of its former territory and population in the Versailles (Trianon) treaties. At the same time, it became a much more homogeneous small country of some 10 million people. The capital city Budapest retained its million inhabitants, acting not unlike Vienna as an oversized head for a much smaller body. Budapest also maintained an unusual concentration of industry and centralized bureaucracy and remained the cultural and educational center of the country.

My mother around 1943

Budapest between the two World Wars was a vibrant, cultured city with excellent theaters, concert halls, opera house, museums, etc. The city consisted of 10 districts. The working-class industrial outskirts of Pest had their row-houses, whereas the middle-class inner city had quite imposing apartment buildings. The upper classes and aristocracy lived in their villas in the hills of Buda.

Some of the social separation, however, was mitigated in the schools. There were remarkably good schools in Budapest, although as with much else it cannot be said that there was a uniformly high level of education across the country. Much was written about the excellence of the gymnasia of Budapest (German-type composite of middle and high schools), particularly those that were frequented by some later successful scientists, artists, and musicians. For example, many Hungarian-born mathematicians, physicists, and engineers who later attained recognition (Neuman, Karman, Wigner, Szilard, and Teller, among others.) attended the same few schools, where much emphasis was put on mathematics and physics. Winning the national competition in these topics for high school boys (the schools were still gender segregated and not much was ever said about the intellectual achievement of Hungarian girls, who clearly were at least as talented) was considered a predictor of outstanding future careers (Neuman, Teller).

I did not attend any of these schools, and there was, furthermore, no national competition in chemistry. After going to a public elementary school, I attended the Gymnasium of the Piarist Fathers, a Roman Catholic religious teaching order, for 8 years. Emphasis was on broad-based education heavily emphasizing classics, history, languages, liberal arts, and even philosophy. The standards were generally high. Latin and German were compulsory for 8 years and French (as a selective for Greek) for 4 years. During my high school years, at the recommendation of one of my teachers, I got my first teaching experience tutoring a middle school boy who had some difficulties with his grades. I enjoyed it and bought with my first earnings an Omega wristwatch, which I had for a quarter of century (I guess it was a good luck charm through some difficult times). I also took private lessons in French and English. In a small country such as Hungary, much emphasis was put on foreign languages. German was still very much a second language as a remnant of the Austro-Hungarian monarchy, and I was fluent in it from the early childhood.

We also received a solid education in mathematics, as well as some physics and chemistry. Although, among others, the physicist Lorand Eotvos and the chemist George Hevesy (Nobel Prize in 1943) attended the same Piarist School in earlier years, I learned about this only years later and cannot remember that anybody mentioned them as role-models during my school years. Classes were from 8 AM to 1 PM 6 days a week, with extensive home work for the rest of the day. I did well throughout my school years in all subjects, except in physical education, where I remember I could never properly manage rope climbing and some exercises on gymnastic instruments. At the same time I was active and became reasonably good in different sports, including tennis, swimming, rowing, and soccer. In my time as a young faculty member at the Technical University I was the goalkeeper of the chemistry faculty team (a picture of which found its way onto the Swedish Academy Nobel poster published after my Prize).

My wife still teases me on occasion that I always was a stellar student (she uses the more contemporary expression "nerd") and school valedictorian. Studying, however, always came easy for me, and I enjoyed it. I was (and still am) an avid reader. In my formative school years I particularly enjoyed the classics, literature, and history, as well as, later on, philosophy. I believe obtaining a good general liberal education was a great advantage, because getting attached too early to a specific field or science frequently short-changes a balanced broad education. Although

Goalie of the soccer team of the chemistry faculty at the Technical University of Budapest, 1952

reading the classics in Latin at school age may not have been as fulfilling as it would have been at a more mature age, once I got interested in science I could hardly afford the time for such pleasurable diversions.

Not yet thinking about chemistry

Besides the classics, Hungarian literature itself offered a wonderful treasure house. It is regrettable that the works of many highly talented Hungarian poets and writers remained mostly inaccessible to the rest of the world because of the rather strange language in which they were written (Hungarian belongs to the so-called Finn-Uogric family of languages, not related to any of the major language families). On the other hand, Western authors and poets were regularly translated into Hungarian, some of these translations being of extremely high quality. The Hungarian poet Janos Arany, for example, translated Shakespeare in a masterly way, although he taught English himself phonetically and could never speak it. I believe that many translations, such as ballads of Villon (one of my favorites), translated by the highly talented poet George Faludy, made them at least as enjoyable as the originals. Operas were sung only in Hungarian, which limited the chances for some talented Hungarian singers to perform in the West.

Western movies were subtitled, but an active domestic movie industry also existed. An indication of the talented people in this industry is reflected by the subsequent role Hungarian émigrés played in the movie industry of the United States and England. These were not limited to movie producers and executives (Adolf Zukor, Paramount Pictures; William Fox, 20th Century Fox; Sir Alexander Korda). Who would guess, considering the thickly accented English most Hungarians speak, that Leslie Howard's (Professor Higgins in *Pygmalion*) mother tongue was this rather obscure and difficult language? What a difference indeed from the charming accent of Zsa Zsa Gabor. On the wall of Zukor's office there was an inscription: "It is not enough to be Hungarian, you must be talented, too." Zukor was quoted to add in a low voice, "But it may help."

The standards of theatrical and musical life were also extremely high. Franz Liszt (a native Hungarian who however never spoke the language) who achieved his fame as a virtuoso pianist and composer established a Music Conservatory in Budapest around 1870, which turned out scores of highly talented graduates including such composers as Kodaly and Bartok and conductors such as Doraty, Ormandy, Solti, Szell, and others.

Good teachers can have a great influence on their students. As I mentioned, besides the classics, humanities, history, and languages I also received a good education in mathematics and had some inspiring science teachers. I particularly remember my physics teacher, Jozsef Öveges, who I understand later, introduced the first television science programs in Hungary, which made him well known and extremely popular. He was a very inspiring teacher who used simple but ingenious experiments to liven up his lectures. I must confess that I do not remember my chemistry teacher, who must have made less of an impression on me. When a friend received a chemistry set for Christmas one year, we started some experiments in the basement of their home. This was when I probably first experienced the excitement of some of the magic of chemistry. For example, we observed with fascination bubbles of carbon dioxide evolving when we dropped some sodium bicarbonate into vinegar or muriatic acid. Little did I expect to repeat this experiment with superacidic "magic acid" years later, when, unexpectedly, no bubbles were formed (see Chapter 8). After going through some of the routine experiments described in the manual that came with the chemistry set, we boldly advanced to more interesting "individual" experiments. The result was an explosion followed by a minor fire and much smoke, which destroyed the chemistry set and ended the welcome for our experimentation. I cannot remember to have thought much about chemistry thereafter, until I entered university.

The outbreak of World War II in September 1939 initially affected our life little. Hungary stayed out of the war until Hitler's invasion of the Soviet Union in June 1941, when it was pressured to join Germany with disastrous consequences. In 1943, a whole Hungarian army corps was destroyed in Russia. Hungary, not unlike Romania, subsequently tried to quit the war. In the spring of 1944, Hungary attempted to break the alliance with Germany but was unsuccessful, and the Gestapo with their Hungarian allies, the Arrow Crosses, took over and instituted a regime of terror. Much has been written about this horrible period when hundreds of thousands perished until World War II finally ended. The human mind mercifully is selective, and eventually we tend to remember only the more pleasant memories. I do not want to relive here in any detail some of my very difficult, even horrifying, experiences of this period, hiding out the last months of the war in Budapest. Suffice it to say that my parents and I survived. My brother Peter (3 years my senior), however, perished at the end of World War II in a Russian prisoner of war camp, where he was taken together with tens of thousands of civilians only to satisfy claimed numbers of prisoners of the war. My home town, beautiful but badly ravaged Budapest, also survived and started recovering. However, soon again difficult times followed when a brief period of emerging democracy ended with the Communist takeover.

I graduated high school (gymnasium) right at the end of the war in the spring of 1945. We needed to pass a very tough 2-day examination (matriculation or baccalaureate). This included, for example, translation of selected literature passages—without the help of a dictionary—not only from foreign languages (including Latin) into Hungarian, but also vice versa, not a minor effort. Having passed this barrier, it was time to think seriously about my future. It was clear that my interests in the humanities, particularly literature and history, did not offer much future in postwar Hungary.

My brother Peter in 1943

One inevitably was forced to think about more "practical" occupations. I had no interest in business, law, or medicine but was increasingly attracted to the sciences, which I always felt in a way to be closely related to the arts and humanities. My eventual choice of chemistry was thus not really out of character. I was attracted to chemistry probably more than anything else by its broad scope and the opportunities it seemed to offer. On one hand, chemistry is the key to understanding the fundamentals of biological processes and may be life itself. It is also an important basis of the life and health sciences. On the other hand, chemists make compounds, including the man-made materials, pharmaceuticals, dyes, fuels, etc., essential to everyday life. It also seemed to be a field in which, even in a poor, small country, one could find opportunities for the future.

My choice of chemistry as a career from a practical point of view was also not unusual. I read years later that Eugene Wigner, the Hungarian-born physicist, when he was 17, was asked by his father (a businessman) what he wanted to do. The young Wigner replied that he had in mind to become a theoretical physicist. "And exactly how many jobs are there in Hungary for theoretical physicists?" his father inquired.

"Perhaps three," replied the young Wigner. The discussion ended there and Wigner was persuaded to pursue chemical engineering. Similarly, the mathematician John von Neuman, initially also studied chemistry, although they both soon changed to their field of real interest.

I entered the Technical University of Budapest immediately after the end of World War II. It had been established some 150 years earlier as the Budapest Institute of Technology. Because the Austro-Hungarian education system closely followed the German example, the Technical University of Budapest developed along the lines of German technical universities. Many noted scientists of Hungarian origin such as Michael Polanyi, Denis Gabor (Nobel Prize in Physics, 1971), George Hevesy (Nobel Prize in Chemistry, 1943), Leo Szilard, Theodor von Karman, Eugene Wigner (Nobel Prize in Physics, 1963), John von Neuman, Edward Teller, and others have studied there at one time or other, generally, however, only for short periods of time. Most of them completed their studies in Germany or Switzerland generally in physics or mathematics, and their career carried them subsequently to England or the United States. To my knowledge, I am the only one who got my entire university education at that institution and who subsequently was also on its faculty. I was proud to be recognized as such when my alma mater welcomed me back after many years in 1988 and gave me an honorary Doctor of Science degree.

Formally, chemistry was taught at the Technical University under the Faculty of Chemical Engineering, but it was really a solid chemical education (along the lines of related German institutions), with a minimum of engineering courses, in sharp contrast to what American universities teach as chemical engineering. Classes were relatively small. We initially started with a class of about 80, but the number was rapidly pared down during the first year to half by rather demanding "do or die" examinations (including comprehensive oral examinations). Those who failed could not continue. This was a rather harsh process, but laboratory facilities were so limited that only a few could be accommodated. The laboratory training however was thorough. For example, in the organic laboratory we did some 40 synthetic preparations (based on Gatterman's book). It certainly gave a solid foundation.

In contrast to the excellent general education some of the gymnasia of Budapest provided their students, the chemical education at the

Technical University in retrospect was very one-sided. The emphasis was on memorizing a large amount of data and not so much on fostering understanding and critical probing. Empirical chemistry, including such fields as gravimetric analysis, the recounting of endless compounds and their reactions, etc., was emphasized, with little attention given to the newer trends of chemistry developing in the Western countries. However, I was well prepared for self-study and gained much through long hours in the library, which opened up many new windows. There were also, however, other very positive aspects of my university education. Many of our professors through their lectures were able to induce in us a real fascination and love for chemistry, which I consider as the most important heritage of my university days.

After completing my studies and thesis, in June 1949, I was named to a junior faculty position as an assistant professor in Zemplen's organic chemical institute. The following month I married my boyhood love.

I met Judy Lengyel, a shy girl probably not much interested in boys, in 1943 during a vacation at a summer resort not far from Budapest.

The shy girl Judy Lengyel

I was 16, and she was 14. Little did I realize at the time that this would become the most significant and happiest event of my life. Although fairytale things are not supposed to happen in real life, we were married 6 years later and celebrated our golden wedding anniversary in 1999. If anybody was blessed with a happy marriage and life partnership, I am. There are, however, no words that can express it adequately.

As newlyweds in 1950

At the time of our marriage, Judy worked as a secretary at the Technical University. She subsequently studied chemistry and after completing her studies, joined our research. To this day, however, I am not sure that she has completely forgiven me for what she probably rightly recollects was my enrolling her to study chemistry (taking advantage of my faculty position) without really getting her prior full approval. Even if I may have been guilty in this regard, my intentions were good. The life of any scientist is very time demanding. If marriage means a real partnership, having common understanding and interest also in our professional life can help to make, or so I hoped, a marriage even more successful. I believe it worked out this way, but only through Judy's efforts and understanding. I can only hope that she indeed has forgiven me after all these years.

Under the oars held up by members of my rowing club after our wedding on July 9, 1949

· 5 ·

Early Research and Teaching:
Departing the Shadow of Emil Fischer

I started my research in Budapest in the Organic Chemistry Institute of Professor Geza Zemplen, a noted carbohydrate chemist of his time and a student of Emil Fischer. Fischer was the towering organic chemist of the early twentieth century. His work, including that on amino acids and proteins, carbohydrates, the start of enzyme chemistry (his famous "lock and key" concept) and many other significant contributions, also laid the foundation for the explosive development of what became biochemistry and, in more recent times, molecular biology. He received the Nobel Prize in chemistry in 1902, the second year the prize was awarded. His students spread across not only Germany but the whole of Europe and established in their home lands the traditions of the Fischer-school. They also contributed much to the study of nature's fundamental systems and their understanding. This was also the true of Zemplen, who established in Hungary the first organic chemistry institute when he was named to a newly established chair in organic chemistry at the Technical University of Budapest in 1913. His work not only resulted in elevating the educational standards of his university together with the development of successful and productive research but also helped to build a successful and respected pharmaceutical chemical industry in Hungary, which continues to the present day.

A Life of Magic Chemistry: Autobiographical Reflections Including Post-Nobel Prize Years and the Methanol Economy, Second Updated Edition. George A. Olah with Thomas Mathew.
© 2015 John Wiley & Sons, Inc. Published 2015 by John Wiley & Sons, Inc.

Having brought home from Berlin Fischer's traditions and an over-riding interest in natural products, particularly carbohydrate and glyco-side chemistry, Zemplen established and ran his laboratory very much in the Fischer style, adding, however, his personal touches. The laboratories had tall ceilings and large windows (not unlike Fischer laboratories which I have seen in the early 1950s at the Humboldt University in Berlin to provide better air circulation because Fischer suffered from chronic hydrazine poisoning from his research). The fume hoods were vented by a gas burner providing draft through the heated air, certainly a not very efficient system by today's standards.

Zemplen, like Fischer, also ran his Institute in an autocratic way. This was even more emphasized by the fact that, because the University was able to provide only an extremely meager budget, laboratory equipment, chemicals, books, etc. were purchased by his own earnings and consulting income. Like Fischer, Zemplen too expected his doctoral students to pay their own way. Fischer even charged them a substantial fee for the privilege of working in his laboratory. Becoming a research assistant to the professor was a great privilege, and, although it meant no remuneration, it also exempted one from any fees.

Zemplen was also a formidable character, and working for him was quite an experience, not only in chemistry. He liked, for example, "pub-bing," and these events in neighborhood establishments could last for days. Certainly one's stamina and alcohol tolerance developed through these experiences. Recalling on occasion his Berlin days, Zemplen talked with great fondness of his laboratory mate the Finnish chemist Komppa (of later camphene synthesis fame) who he credited not only with being a fine chemist but also with being the only one able to outlast him during such parties. In any case, none of this ever affected his university duties or his research.

Zemplen helped his students in many ways. I remember an occasion in the difficult postwar period. The production of the famous Hungarian salami, interrupted by the war, was just in the process of being restarted for export. The manufacturer wanted a supportive "analysis" from the well-known professor. Zemplen asked for a "suitable" sample of some hundreds of kilograms, on which the whole institute lived for weeks. When it was gone he rightly could write his opinion that the product was "quite satisfactory." After the war, grain alcohol was for a long time the only available and widely used laboratory solvent, and, not unexpectedly, it also found other uses. Later, when it was "denatured" to prevent human consumption, we devised clever ways for its "purification." The laboratory, for a while, also "manufactured" saccharine, which was then bartered for essential food staples in the countryside. One of my colleagues suffered serious burns one day, when a flask containing chlorosulfuric acid broke in a water-ice bath, spraying acid into his face. Such was the price of moonlighting for survival.

Zemplen was a strong-minded individualist who opposed any totalitarian system, from Nazism to Communism. He was briefly jailed toward the end of World War II by the Hungarian Fascists for refusing to join in the evacuation of the Technical University to Germany when the Russian armies advanced on Budapest. He was also strongly opposed to the Communists.

As Zemplen was a student of Emil Fischer, therefore, I can consider myself Fischer's "scientific grandson" as shown in a memorial published on the centennial of his Nobel Prize. My initial research with Zemplen was along the lines of his interest in natural products and specifically centered on glycosides. Besides synthetic and some structural work, there also emerged practical aspects. Many glycosides present in nature have substantial pharmaceutical use. For example, some of the most effective heart medications were isolated from the pretty foxglove plant (*Digitalis purpurea*). Three cardiac glucosides had been isolated by the Swiss chemist Stoll, working for the Sandoz Company (digitoxin, gitoxin, and gitalin). Another member of the *Digitalis* family, *Digitalis lanata*, contains lanataglucoside C, whose aglucone is also digitoxin (still one of the most effective heart medication). As it turned out, the mild climate of the Tihany peninsula, jutting into famed Lake Balaton, favored the cultivation of this plant, and each year boxcar loads of its leaves were shipped to Sandoz in Switzerland. While working with Zemplen, I developed an improved process for the efficient isolation of lanataglycoside C, using treatment and solvent extraction of fresh leaves suppressing enzymatic degradation. This greatly increased yields, and a gram of pure lanataglycoside C could be obtained from a kilogram of leaves. Our process with Zemplen was patented in 1952 and, together with a process for partial hydrolysis to digitoxin, was successfully commercialized and used for years by the Richter Pharmaceutical

Company. A booklet commemorating the 100th anniversary of the Hungarian Patent Office mentioned our patent as well as other Hungarian patents, such as those issued in the 1920s to Albert Einstein and Leo Szilard for a thermoelectric refrigerator (which, I understand, regrettably never brought any income to the inventors although they even established a company in Berlin). In the 1930s in England, Szilard, the remarkable Hungarian-born physicist, applied for a patent on another invention, the principle of a thermonuclear device, that is, the basis of the atomic bomb! In 1952, he and Enrico Fermi were issued an American patent for it. However by law interim passed, the US Government acquired it for the required minimum remuneration of one dollar. What wide scope some inventors' creative mind cover!

I have fond memories of my brief period as a natural product chemist, particularly because it de facto still involved collecting and isolating products from nature's diverse plant life and doing chemistry directly on them. Most natural product chemists these days do not have this experience.

Around 1950, after having read about organic fluorine compounds, I became interested in them. I suggested to Zemplen that fluorine-containing carbohydrates might be of interest in glycoside-forming coupling reactions. I believed that selective synthesis of α- or β-glycosides could be achieved by treating either acetofluoroglucose (or other fluorinated carbohydrates) or their relatively stable, deacetylated free fluorohydrins with the appropriate aglucons. His reaction to my suggestion was, not unexpectedly, negative. To try to pursue organofluorine chemistry in postwar Hungary was indeed far-fetched. Zemplen thought that the study of fluorine compounds, which necessitated "outrageous" reagents such as hydrogen fluoride, was foolish. It became increasingly clear that my ideas and interests did not match his. Eventually, however, I prevailed and, with some of my early dedicated associates, including Attila Pavlath (who, after a career in industry and governmental laboratory research, had been elected for a term the President of the American Chemical Society) and Steve Kuhn (a recently deceased longtime researcher who worked with me at Dow Chemical Co.), we started to study organic fluorides. Laboratory space and particularly fume hoods were scarce in the Zemplen laboratories, and even when I became an assistant professor I was not welcome to "pollute" space intended for "real chemistry." However, at the top floor of the chemistry building, overlooking the Danube and badly damaged during the war, was an open balcony used to store chemicals. In an unexpected gesture, Zemplen allowed me the use of this balcony. With some effort we enclosed it, installed two old fume hoods and were soon in business in what was referred to as the "balcon laboratory." I am not sure that Zemplen ever set foot in it. We, however, enjoyed our modest new quarters and the implicit understanding that our fluorine chemistry and subsequent study of Friedel–Crafts reactions and their intermediates was now approved.

Carrying out fluorine research in postwar Hungary was clearly not easy. There was no access even to such basic chemicals as anhydrous HF, FSO_3H, or BF_3, and we had to prepare them ourselves. HF was prepared from fluorspar (CaF_2) and sulfuric acid, and its reaction with SO_3 (generated from oleum) gave FSO_3H. By treating boric acid with fluorosulfuric acid we made BF_3. The handling of these chemicals and their use in a laboratory equipped with the barest of necessities was indeed a challenge. Zemplen's consent to our work was even more remarkable, because he truly belonged to the "old school" of professors who did not easily change their mind. For example, he never believed in the electronic theory of organic chemistry. In 1952, he was persuaded to publish a Hungarian textbook of "Organic Chemistry," which contained a wealth of information and discussion of natural products but otherwise still treated organic chemistry only in the descriptive ways of the nineteenth and early twentieth centuries. An interesting side aspect of this book was that it had a very substantial index, for which, with a faculty colleague (Lajos Kisfaludy), we were commissioned by the state publishing house on a per page basis. Our rather precarious financial existence (our meager university salary amounted to the equivalent of less than $50/month, which was gradually gone by the middle of the month) inspired us to a remarkable effort. Our index eventually contained 250 pages (a fifth of the book) and even displayed the preparers' names. I believe it is still the longest index of any textbook of chemistry. It had, however, seen us through some difficult times.

No mechanistic aspects of organic chemistry (or, for this reason, any reaction intermediates) were ever mentioned by Zemplen in his lectures or writings, nor did he consider or accept their existence. I never heard him mention the names of Meerwein, Ingold, Robinson, or any other pioneers of the mechanistic electronic theory of organic chemistry. The possible role of organic ions was similarly never mentioned. He was, however, not alone with this attitude at that time. European, particularly German, chemistry generally was slow to accept Robinson and Ingold's electronic–mechanistic concepts. Even Roger Adams, one of the leading American organic chemists of the time (and a descendent of the family that gave two presidents to America), also, never believed in organic ions. For a long time his influence prevented anybody including Whitmore, who pioneered the concept of the role of cationic intermediates of organic reactions from publishing any article in the *Journal of the American Chemical Society or any other journal* of mentioning such ions. Clearly, working on such problems in the Zemplen Institute came close to sacrilege, but I somehow got away with it. Perhaps this was in part due to the fact that Zemplen's health was declining at the time in his long battle with cancer. He came less frequently to the Institute, working mostly at his nearby apartment.

Did the Technical University in Budapest at this time provide a reasonable education and research atmosphere in chemistry? In retrospect, it is difficult to answer this question because of the prevailing isolation and primitive working conditions. The scope and topics of major courses taught, with some exceptions, mainly demanded memorization of facts and not understanding and discussion of concepts. Laboratory experimental work and research were much emphasized, despite the very limited facilities and opportunities. The overall system, however, prepared students well for self-study. I myself discovered at an early age the joy of broadening my scope and knowledge during long hours spent in the library in extensive reading and study (a habit I have kept all my life). Senior professors always lectured themselves, which was considered a major prerequisite of the professorship. Attendance by all students was strictly obligatory (the system was tailored on the German example). Lectures particularly by popular professors were considered worthy and were well-attended events. I remember also attending many lectures outside my field in the Budapest universities such as by philosophers, historians, etc., which were always packed to the rafters with interested students.

In chemistry clearly there was a great need to move ahead and bridge the gap between the earlier, entirely empirical approach of teaching and research and that incorporating the new areas of our knowledge. The emerging understanding of mechanistic–structural aspects of chemistry, application of physical methods and considerations, and the basic principles of bonding, reactivity, etc., were rarely mentioned before the 1950s. When I had the chance starting in 1953 to teach my own course, it was entitled "Theoretical Organic Chemistry" (although really it was mechanistic and structural, i.e., physical organic chemistry). It filled a gap in which I tried to introduce the evolving new frontiers of organic chemistry. I also published a two-volume text based on the notes of my two-semester course in 1953–1954. An extended and reworked first volume written with contribution by Attila Pavlath was submitted to the publisher Verlag Chemie in 1956 but was published in German only in 1960 (G. Olah, *Theoretische Organische Chemie*, Akademie Verlag, Berlin, 1960) because of events that led to my leaving Hungary. The second volume of the German edition was however never published. At that time in Germany, physical organic chemistry was also slow to develop. There was an early prewar text by Eugen Müller, but it was Heinz Staab's book in the 1950s that opened up the field. Looking back, I realize that despite my isolation beyond the Iron Curtain, I still somehow managed to get myself oriented in the right direction. My book at the time was reasonably up to date and in some aspects may be even novel and original.

During the time of teaching on the faculty of the Technical University I was also involved in helping to start a technical high school associated with the university. This school filled a gap to allow evening study for students who did not have the opportunity for usual high school education. It provided solid secondary education with emphasis on the sciences and technology, qualifying its students for meaningful employment as technicians and in other technically oriented jobs. I remember that I even taught physics for a year, when we had difficulty finding a teacher for physics. To my surprise, I managed quite well.

My teaching experience was, however, only secondary to my research interest. Through my initial research work involving reactions of fluorinated carbohydrates I became interested in Friedel–Crafts acylation and subsequently alkylation reactions with acyl or alkyl fluorides, with boron trifluoride used to catalyze the reactions. This also was the beginning of my long-standing interest in electrophilic substitution reactions.

$$\text{ArH + RCOF} \xrightarrow{\text{BF}_3} \text{ArCOR + HF}$$

$$\text{ArH + RF} \xrightarrow{\text{BF}_3} \text{ArR + HF}$$

These studies at the same time aroused my interest in the mechanistic aspects of the reactions, including the complexes of RCOF and RF with BF_3 (and eventually with other Lewis acid fluorides) as well as the complexes they formed with aromatics. I isolated for the first time at low temperatures arenium tetrafluoroborates (the elusive σ-complexes of aromatic substitutions), although I had no means to extensively study their structural aspects. Thus my long fascination with the chemistry of carbocationic complexes began.

$$\text{ArH + HF + BF}_3 \rightleftharpoons \text{ArH}_2^+ \text{BF}_4^-$$

$$\text{ArH + RF + BF}_3 \rightleftharpoons \text{ArHR}^+ \text{BF}_4^-$$

My early work with acyl fluorides also involved formyl fluoride, HCOF, the only stable acyl halide of formic acid, which was first made in 1933 by Nyesmeyanov, who did not, however, pursue its chemistry. I developed its use as a formylating agent and also explored formylation reactions with CO and HF, catalyzed by BF_3 which later became industrially used.

Another field of my early research in Budapest was nitration chemistry using specifically nitronium tetrafluoroborate, a stable nitronium salt. I was able to prepare the salt in a simple and efficient way from nitric acid, hydrogen fluoride, and boron trifluoride.

$$\text{HNO}_3 + \text{HF} + 2\,\text{BF}_3 \longrightarrow \text{NO}_2^+\text{BF}_4^- + \text{BF}_3 \cdot \text{H}_2\text{O}$$

This salt turned out to be remarkably stable and a powerful, convenient nitrating agent for a wide variety of aromatics (and later also aliphatics). Over the years, this chemistry was further developed, and nitronium tetrafluoroborate is still a frequently used commercially available nitrating agent.

$$\text{ArH} + \text{NO}_2^+ \text{BF}_4^- \longrightarrow \text{ArNO}_2$$

$$\text{RNH}_2 + \text{NO}_2^+ \text{BF}_4^- \longrightarrow \text{RNHNO}_2$$

$$\text{ROH} + \text{NO}_2^+ \text{BF}_4^- \longrightarrow \text{RONO}_2$$

In the course of my studies I also introduced silver tetrafluoroborate, $AgBF_4$, as a metathetic cation forming agent suitable for forming varied ionic (electrophilic) reagents.

$$\text{RCl} + \text{AgBF}_4 \longrightarrow \text{R}^+\text{BF}_4^- + \text{AgCl}$$

My publications from Hungary in the early 1950s somehow caught the eye of Hans Meerwein, a towering and pioneering German chemist who extended the chemistry of $AgBF_4$. I received an encouraging letter from him, and we started corresponding (even this was not easy at the time in isolated Hungary). He must have sympathized with my difficulties, because one day through his efforts I received as a gift a cylinder of boron trifluoride from BASF. What a precious gift it was, because it freed us from laboriously making BF_3 ourselves.

Hans Meerwein, always an inspiration

My early research on organofluorine compounds starting in 1950 centered on new methods of their preparation and use in varied reactions. At the time I also started a collaboration with Professor C. Sellei, a wonderful physician associated with the Medical University of Budapest. We studied the pharmacological effect of organofluorine compounds and particularly on the use of some organic fluorine compounds in cancer research. Sellei and his researcher wife Gabi became close friends and the godparents of our elder son, George. Our joint publications from 1952 to 1953 must be among the earliest ones in this intriguing field. In subsequent years, the work of Charles Heidelberger (who became a colleague when I moved later to the University of Southern California in Los Angeles) and other researchers introduced such widely used fluorinated anticancer drugs as 5-fluorouracil. The pursuit of the biological activity of fluorinated organic compounds clearly was a worthwhile early effort, as progress in this field is still remarkable.

At the time, when I was fascinated by the potential of organofluorine compound-based cancer research, I decided to learn more about biological and medicinal aspects. While maintaining my university appointment and duties, I enrolled in the Medical University of Budapest. I passed all the preclinical courses, including the rather challenging anatomy course, but never really intended to obtain an MD. What I learned, however, served me well, because it is difficult to appreciate the biomedical field and its revolutionary expansion without some proper foundations.

The Hungarian educational system after the Communist takeover was realigned in the early 1950s according to the Soviet example. University research was de-emphasized, and research institutes were established under the auspices of the Academy of Sciences. I was invited in 1954 to join the newly established Central Chemical Research Institute of the Hungarian Academy of Sciences, which was headed by Professor Geza Schay, a physicochemist with primary interest in surface chemistry and catalysis. He was a warm, truly fine man, who considered his main goal to help the development of his younger colleagues. I was honored that he chose me to become his deputy in our quite modest Institute. I established a small organic chemistry research group with six to eight members, which also included my wife. Our laboratories were in rented space in an industrial research institute on the outskirts of Pest, but I certainly had more space than my original small "balcon laboratory" at the Technical University. Conditions, however, were still difficult. I never would have thought that this small research institute eventually would grow after my departure from Hungary into a large institution with hundreds of researchers, which is what I found after more than 25 years when I revisited my native Hungary for the first time. In the post-Communist era, however, university research became again encouraged and the oversized research institutes of the Academy of Sciences are slowly being cut back to a more reasonable size or merged.

We were very much isolated in Hungary from the scientific world, which is the worst thing that can happen to scientists. Journals and books were difficult to obtain, and even then only with great delays. I was able to attend only one scientific meeting in the West, the 1955 meeting of the International Union of Pure and Applied Chemistry (IUPAC) in Zurich, where I gave a lecture about some of our research. It was a unique experience to meet for the first time many chemists whose work I knew only from the literature. Little did I foresee that a year later my life was going to change drastically and such contacts would regularly become possible.

Contacts even with colleagues in the Soviet Union and other Eastern countries were also difficult. In 1955–1956, I gave some lectures in what was then East Germany, including in East-Berlin at the Humboldt University, where I was able to see Emil Fischer's old laboratories. I also took the opportunity to cross into West-Berlin (this was before the wall had gone up) to visit some of the organic chemists (Weygand, Bohlmann). At an East German scientific meeting I met some fine Czech colleagues, among others, Lukes (whose former graduate student was Vlado Prelog, Nobel Prize in 1975) and Wichterle (the later inventor of soft contact lenses) a very brave man who had a significant part in the Czech Velvet-Revolution in the late 1960s. My closest contact was with Costin Nenitzescu, an outstanding chemist from Bucharest, Rumania. He visited Budapest on some occasions. We met again in the late 1960s when he visited me in Cleveland and we kept in close touch until his untimely death. Interestingly, even contacts with the Soviet Union were weak. Remarkably, information about Russian research at the time was easiest to obtain from the *Chemical Abstracts* published by the American Chemical Society. At the East German meeting I mentioned, I met Professor Reutov and eventually through him Nesmeyanov. He not only was a leading chemist but also a great power in Soviet science, being the president of the Academy of Sciences, which had a formidable string of large research institutes. I visited the Soviet Union only once, accompanying in 1955 Geza Schay, the director of our research institute. Because Nesmeyanov was interested in some of my work in areas where he himself had worked, when we were in Moscow I was taken to see him in his Institute of Organoelement Compounds, a large and impressive place with many hundreds of researchers. We had a good discussion in his office. We spoke French and German (I do not speak Russian), but, surprisingly, when somebody came in to join us, he changed to Russian and asked for an interpreter. Even he seemed to be concerned at the time not to be reported for such "anti-Soviet" behavior. In our private conversation, Nesmeyanov told me about the recent condemnation of Pauling's resonance theory. He mentioned that he had not signed the official report orchestrated by the political powers, claiming illness so as not to participate in the process. Even in retrospect, it showed a most discouraging picture of the political influence on Soviet science.

Other memories I have about my visit include meeting Kitaigorodsky, an excellent X-ray crystallographer. He opened the door to a large room filled with tables at which set row upon row of girls cranking mechanical calculators. With a sad smile, he said that this was his computer, but he had three shifts of girls working at the calculators. The situation was different at Semenov's Institute of Chemical Physics (he received the Nobel Prize, shared with Norrish, in the following year, 1956). He told us in some detail inter alia about his state-of-the-art work on shock waves and high-energy propulsion systems. I realized only after Sputnik was launched in 1957 that his work must have had been connected with it.

Isolation clearly was a most depressing aspect of pursuing science in Communist-dominated Hungary. To overcome it, sometimes one even tried to attempt unusual paths. In 1955, I read in a scientific journal an announcement about the Dutch Van't Hoff Fellowship, which offered to the winner a visit to Holland to meet with Dutch scientists. Although

I had no permission (it may seem strange, but at the time we were not allowed even such contacts with the West), I applied for it, not expecting to hear anything further about it. To my great surprise, however, I was chosen for the fellowship in early 1956, but I was not given the opportunity at the time to visit Holland. However, in 1957, by which time I lived in Canada, I was able to lecture at an IUPAC conference in Paris and, on my way, visit Holland. I greatly enjoyed meeting Dutch colleagues, particularly Professors Wibaut and Sixma, and enjoyed the gracious hospitality offered to me.

I spent two very productive years in the Chemical Research Institute of the Hungarian Academy of Sciences until the fall of 1956, making the best use of existing research possibilities. In October 1956, Hungary revolted against Soviet rule, but the uprising was soon put down by drastic measures after intense fighting with overwhelming Soviet forces and much loss of life. Budapest was again devastated, and the future looked bleak. Our research institute was not itself damaged during the fighting, but our spirit was broken. During the short-lived days of freedom, a small "revolutionary committee" was formed in our Institute to which I was elected. There was nearly unanimous support for the spontaneous revolt of 1956 in Budapest, but it could not on its own prevent the oppression and terror that returned with the Soviet forces. Consequently, in November and early December of 1956 some 200,000 Hungarians, mostly of the younger generation, fled their homeland. With my family (our older son was born in 1954) and most of my research group, who also decided that this was the only path to follow, we joined the torrent of refugees seeking a new life in the West.

· 6 ·

Move to North America:
Industrial Experience While Pursuing
the Elusive Cations of Carbon

After fleeing Hungary with most of my research group in the exodus of November of 1956, in early December with my wife, and our 2-year-old son, we eventually got to London, where an aunt of my wife lived. We were warmly welcomed. During our stay in London I was able to meet for the first time some of the chemists whose work I knew from the literature and admired. I found them most gracious and helpful. In particular, Christopher Ingold and Alexander Todd (Nobel Prize in chemistry 1957) extended their efforts on behalf of a young, little-known Hungarian refugee chemist in a way that I will never forget and for which I will always be grateful.

While in London I visited Ingold at University College and also met some of his colleagues, including Ed Hughes, Ron Nyholm, and Ron Gillespie (with whom I later renewed contact in Canada). I was invited in early January 1957 to give a seminar at Cambridge. This was the first time I ever lectured in English. My talk raised some comments (which I understand were made before in connection with another Hungarian émigré's similar experience) on "how interesting it is that this strange Hungarian language has some words which resemble English." Anyhow, I survived my "baptism," but, like many Hungarian-born scientists, I retain to this day an unmistakable accent (although some believe that this may be an asset). I remember about my visit that I was picked up at the railroad station by my host Bob Haszeldine (whose work I knew through my work in fluorine chemistry), who was driving what turned out to be a pricey Bentley. He was a lecturer at the time and was helping Alexander Todd (later the Nobel Laureate Lord Todd) with the design of the new Lensfield Road chemistry building. He moved subsequently to Manchester and had a major part in building the Manchester Institute of Technology. In late afternoon, I was dropped off at the same station by Todd in his small Morris. I did not yet understand the significance some in Western society attach to the kind of the car they drive and its assumed social implication; Todd certainly was not one of them. Another memory of my visit was a conversation in which Al Katritzky, another young member of the faculty who was preparing to join the planned new "red brick" University of East Anglia, was telling Todd about his plans for the new chemistry department. They included even a number of NMR spectrometers, quite extravagant at the time. Todd listened politely and then remarked: "if your plans work out properly, in a few centuries you indeed may have a good university."

A heartwarming experience for my wife and myself was meeting Ms. Esther Simpson, who was running the Academic Assistance Counsel (AAC) and had a modest office at Piccadilly next to Burlington House. She wrote many letters on my behalf trying to help with my search for a job and our resettling. She also extended to us constant kindness and encouragement. I learned only years later that with Leo Szilard she was instrumental in helping to organize AAC as a clearinghouse to match refugees of the Hitler era with placement possibilities. Ms. Simpson was the essential dynamo of the organization and helped many refugee scientists. My own situation was probably not different from that faced by many others over the years who learned to appreciate her warm humanity.

We did not intend to settle in England and were looking forward to moving to Canada; my mother-in-law lived in Montreal, having remarried there after the war. Todd, having learned about our plans to move to Canada, strongly recommended that I should go to Saskatoon, Saskatchewan, where he had a nephew, a professor in the medical school, whom he visited on occasion. He felt the university there was a pleasant, promising place. When I looked it up, however, it looked very much to be in the middle of the prairie, wide-open as far as the North Pole. It turned out that Todd's visits were always during the summer. Having learned more about the long, harsh winters of Saskatoon, we were discouraged from following his advice. (I learned later that Gerhardt Herzberg spent years in Saskatoon before moving to Ottawa, but he was probably a sturdier person.) In March 1957, we traveled by air to Montreal, and I renewed my efforts of looking for a job to be able to support my family. I felt that perhaps in my native small Hungary I was already an established researcher with some recognized achievements, but I learned fast that in the new world this meant little. Canada at the time (perhaps not much differently from the present) had few research opportunities. At Ingold and Todd's recommendation, Maurice Steacie, who was heading the National Research Council in Ottawa, offered me a postdoctoral fellowship after a visit there, but I needed a permanent job. While I was initially looking for an academic position, none came along, but a few industrial research possibilities opened up. Of these, that at Dow Chemical was the most interesting.

Dow Chemical, with its home base at Midland, Michigan, was establishing at the time a small exploratory research laboratory 100 miles across the Canadian border in Sarnia, Ontario, where its Canadian subsidiaries' major operations were located. I was offered the opportunity to join this new laboratory, and two of my Hungarian collaborators who also came to Canada, including Steve Kuhn, could also join me. We moved to Sarnia in May of 1957. As our moving expenses were paid for with a feeling of extravagance, we checked in the two cardboard boxes containing all of our worldly possessions for the train trip and started out for our destination on the shore of Lake Huron, 50 miles north of Detroit. Our younger son, Ronald, was born in Sarnia in 1959, and we have fine memories of the 7 years we spent there. It was a pleasant, small industrial city with a large concentration of petrochemical industries and refineries, but the residential areas were in the north of town, close to the lake and away from the industrial area. The surroundings were well suited to bringing up a young family. There was, however, no possibility for Judy to continue her career in research, because industry at the time would not employ a married couple. However, she rejoined our joint effort when in 1965 we moved to Cleveland and I returned to academic life.

The Olah family, 1962

In the 1950s it was a general trend for major American chemical companies to establish European research laboratories. For example, Cyanamid in Geneva, Union Carbide in Brussels, and Monsanto in Zurich set up such laboratories with impressive staffs of fine chemists such as Hudson, Jorgensen, and Klopman at Cyanamid, Dahl, Schröder, and Viehe at Union Carbide, Zeiss at Monsanto, and others. Dow kept out of Europe for a long time (perhaps because of the lingering effect of its preWorld War II German industrial contacts, which resulted in congressional inquiries during the war). Instead, in 1955 it established an Exploratory Research Laboratory in Canada financed by and responsible directly to the parent company. This small laboratory (with a staff of perhaps 15–20) had a significant degree of freedom and close contact with Dow research in Midland, MI, just 2 hours drive across the border from Sarnia.

As a chemist, it was easy for me to fit into the new environment. Science is international and even has its own language in scientific English (as it is frequently called), that is, English with a foreign accent. The initial language difficulty many immigrants face is certainly much easier to overcome in the technical and scientific fields.

What was more of a challenge was adapting to the ways of an American-style laboratory and its research practices. Young research chemists were paid salaries not significantly higher than those of technicians, who were, however, unionized. Work was consequently carried out mostly by PhD chemists with few technical support personnel (a situation different from European practices of the time). We set up our laboratory and it was possible to restart active research soon after my arrival at Dow. I was grateful and still am for the opportunity and support given to me. I also look back with some envy and nostalgia to this period, when I spent most of my time working at the bench doing experimental work. This, however, still left the evenings and weekends for pursuing my own research interests, for reading, writing, and thinking about new problems and projects.

After renting for a short while, with the help of mother-in-law's life savings, we bought a small house close to Lake Huron quite a distance from the industrial area. We furnished it with all the necessities of our new life, as we literally had nothing when we arrived. For 2 years we had no car, still believing at the time in such old-fashioned principles as not going into debt and paying cash for everything we purchased. This, of course, caused problems in getting around, including going to work, relying on spotty public bus service, or getting rides from friends. When young, however, one notices such inconveniences much less.

I must also have made other impressions on my new colleagues with my work habits. In an industrial laboratory, located within a chemical plant complex, nobody thought of working extra long hours, weekends, etc. except for some "strange" scientists. However, what I did "after hours" was conceded to be research to pursue my own interests. This suited me well, and I enjoyed the opportunity and pursued it vigorously. In the 8 years I spent with Dow, I published some hundred scientific papers (on which I did the good part of the experimental work myself). I also obtained some thirty patents on various topics of industrial interest. Some of my other discoveries, for example, on the beginnings of superacidic systems and related chemistry, however, were not considered worth patenting, and many of my internal discovery memoranda were never developed into patent applications. On my own time I also worked on the updated German version of my book, *Theoretical Organic Chemistry*, which was originally published in Hungary. It was finally published in 1960 in Germany. I also completed editing with Judy's help at my own time (and in no small part writing it) a four-volume comprehensive monograph, *Friedel-Crafts and Related Reactions*, published in 1963–1965 by Wiley–Interscience. This was a major effort, as it comprised close to 4000 pages and more than 10,000 references. Whereas I was subsequently given considerable freedom to pursue my research interests, a good part of my research was also directed to questions and problems related to the company's industrial interests, which of course were not published in the literature.

My years at Dow were productive and rewarding. It was during this period in the late 1950s that my breakthrough work on long-lived stable carbocations, which in a way I had already started back in Hungary, was carried out. Dow was and is a major user of Friedel–Crafts-type chemistry, including the manufacture of ethylbenzene for styrene production by the reaction of benzene with ethylene. This was assumed to involve cationic intermediates (i.e., carbocations), which were, however, never observed or studied. My work thus also had practical significance to some degree and helped to improve the large-scale industrial process. I was treated generously by Dow and was promoted rapidly to Research Scientist, the highest research position without direct administrative responsibility. I also gained much practical experience in the real, industrial world of chemistry, which served me well through the rest of my career, while at the same time continuing my own more academically oriented interests.

During my time at Dow I was also able to establish personal contact with many researchers whose work I previously knew only from the literature. Although our laboratory was located in a small Canadian industrial city off the path academic visitors usually frequent, I succeeded in starting a seminar program. An increasing number of outstanding chemists came, perhaps as a personal favor, to visit our small industrial laboratory, including Christopher Ingold, Georg Wittig, Rolf Huisgen, and others. I much appreciated their kindness. Another rewarding aspect of my years with Dow was that I was given the chance to publish much of my own work and to lecture at different international symposia and meetings, such as the IUPAC Congresses in Paris (1957), Munich (1959), and London (1963), the Cork (Ireland) Conference on Physical Organic Chemistry (1961), and various meetings and symposia of chemical societies. I used these occasions also to visit and give lectures at many universities. My personal contacts and my circle of colleagues I met grew, many of whom over the years became friends.

In the summer of 1963, I learned that I had won the American Chemical Society Award in Petroleum Chemistry for my work on Friedel–Crafts chemistry. It was a most welcome recognition for someone who only a few years earlier had fled his native country and started all over on a far-away continent. Although I have received numerous other awards and recognitions over the years, with the exception of the Nobel Prize, no other awards touched me as much. I remember that my first ACS Award carried with it a check for $5000. My research director for some reason believed that a company employee was not entitled to any external "income" and I should give this to the company. He also raised some questions concerning royalties of the books I was writing in my spare time. The matter was finally settled by the president of the parent company, who found that these questions were of no merit. I was able to keep my award money and modest book royalties (which I believe anyhow would not have made any impact on the bottom line of the company's earnings).

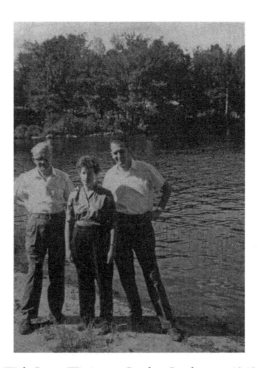

With Georg Wittig at a Gordon Conference, 1960

In my "sailing" outfit, in jacket and tie, with Phil Skell and Heini Zollinger (Cork, Ireland, July 1964)

At London IUPAC congress in 1963

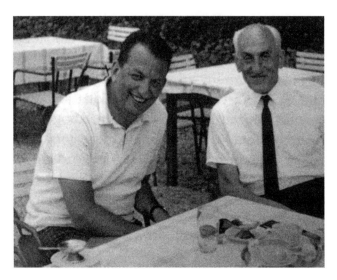

With my father in Austria during a European visit in 1965

Concerning my research during my Dow years I discussed in Chapter 4 that my search for cationic carbon intermediates started back in Hungary while I was studying Friedel–Crafts-type reactions with acyl and subsequently alkyl fluorides catalyzed by boron trifluoride. In the course of these studies I observed (and, in some cases, isolated) intermediate complexes of either donor–acceptor or ionic nature.

$$RCOF + BF_3 \rightleftharpoons \overset{\delta+}{\underset{F}{RCO}}\!\!\rightarrow\!\overset{\delta-}{BF_3} \rightleftharpoons RCO^+ BF_4^-$$

$$RF + BF_3 \rightleftharpoons \overset{\delta+}{RF}\!\!\rightarrow\!\overset{\delta-}{BF_3} \rightleftharpoons R^+ BF_4^-$$

Many elements readily form ionic compounds such as table salt (Na^+Cl^-), in which the cationic sodium and anionic chlorine are held together by electrostatic, ionic bonding. Carbon, however, was long considered to lack the ability to form similar ionic compounds, except in very specific, highly stabilized systems, such as triphenylmethyl dyes.

The Chicago chemist Stieglitz (whose noted photographer brother was the husband of the painter Georgia O'Keefe) suggested in 1899 the possibility of ionic carbon compounds, but it aroused no interest.

When I gave the Stieglitz memorial lecture in the 1980s at the University of Chicago and reminded the audience of Stieglitz's pioneering idea, there was also little appreciation of it. It was in 1901 that Norris and Kehrman, as well as Wentzel, independently discovered that colorless triphenylmethyl alcohol gave deep yellow solutions in concentrated sulfuric acid. Triphenylmethyl chloride similarly formed orange complexes with aluminum and tin chlorides.

Von Baeyer (Nobel Prize, 1905) should be credited for having recognized in 1902 the salt-like character of the compounds formed. He then suggested a correlation between the appearance of color and the formation of salt—the so-called "halochromy." Gomberg (who had just shortly before discovered the related stable triphenylmethyl radical), as well as Walden, contributed to the evolving understanding of the structure of related cationic dyes such as malachite green.

Whereas the existence of ionic triarylmethyl and related dyes was thus established around the turn of the twentieth century, the more general significance of carbocations in chemistry long went unrecognized. Triarylmethyl cations were considered an isolated curiosity of chemistry, not unlike Gomberg's triarylmethyl radicals. Not only were simple aliphatic hydrocarbon cations believed to be unstable, even their fleeting existence was doubted.

One of the most original and significant ideas in organic chemistry was the suggestion by Hans Meerwein that carbocations (as we now call at my suggestion all the positive ions of carbon compounds) might be intermediates in the course of reactions that start from nonionic reactants and lead to nonionic covalent products.

In 1923, Meerwein, while studying the Wagner rearrangement of camphene hydrochloride to isobornyl chloride with van Emster, found that the rate of the reaction increased with the dielectric constant of the solvent. Furthermore, he found that certain Lewis acid chlorides such as

SbCl$_5$, SnCl$_4$, FeCl$_3$, AlCl$_3$, and SbCl$_3$ (but not BCl$_3$ or SiCl$_4$) as well as dry HCl, which promote the ionization of triphenylmethyl chloride by the formation of carbocationic complexes, also considerably accelerated the rearrangement of camphene hydrochloride to isobornyl chloride. Meerwein concluded that the isomerization actually does not proceed by the way of migration of the chlorine atom but by a rearrangement of a cationic intermediate. Hence, the modern concept of carbocationic intermediates was born. Meerwein's views were, however, greeted with much skepticism by his peers in Germany, discouraging him from following up on these studies (see Chapter 9).

camphene hydrochloride

isobornyl chloride

Ingold, Hughes, and their collaborators in England, starting in the late 1920s, carried out detailed kinetic and stereochemical investigations on what became known as nucleophilic substitution at saturated carbon and polar elimination reactions. Their work relating to unimolecular nucleophilic substitution and elimination, called S$_N$1 and E1 reactions, on which the formation of carbocations is the slow rate-determining step, laid the foundation for the role of electron-deficient carbocationic intermediates in organic reactions.

Frank Whitmore in the United States in the 1930s generalized these concepts to include many other organic reactions in a series of papers. Carbocations, however, were generally considered to be unstable and transient (short lived) because they could not be directly observed. Many leading chemists, including Roger Adams, determinedly doubted their existence as real intermediates and strongly opposed even mentioning them. Whitmore, consequently, was never able, in any of his papers in the prestigious *Journal of the American Chemical Society*, to use the notation of ionic R_3C^+. The concept of carbocations, however, slowly grew to maturity through kinetic, stereochemical, and product studies of a wide variety of reactions. Leading investigators such as P. D. Bartlett, C. D. Nenitzescu, S. Winstein, D. J. Cram, M. J. S. Dewar, J. D. Roberts, P. v. R. Schleyer, and others contributed fundamentally to the development of modern carbocation chemistry. The role of carbocations as one of the basic concepts of modern chemistry is well reviewed. With the advancement of mass spectrometry, the existence of gaseous carbocations was proven, but this could not give an indication of their structure or allow extrapolation to solution chemistry. Direct observation and study of stable, long-lived carbocations, such as alkyl cations in the condensed state, remained an elusive goal.

My work on long-lived (persistent) carbocations dates back to the late 1950s at Dow and resulted in the first direct observation of alkyl cations. Subsequently, a wide spectrum of carbocations as long-lived species was studied using antimony pentafluoride as an extremely strong Lewis acid and later using other highly acidic (superacidic) systems.

Until this time alkyl cations were considered only as transient species. Their existence had been indirectly inferred from kinetic and stereochemical studies. No reliable spectroscopic or other physical measurements of simple alkyl cations in solution or in the solid state were obtained, however.

It was not fully realized until my breakthrough using superacids (vide infra) that, to suppress the deprotonation of alkyl cations to olefins and the subsequent formation of complex mixtures by reactions of olefins with alkyl cations, such as alkylation, oligomerization, polymerization, and cyclization, acids much stronger than those known and used in the past were needed.

$$(CH_3)_3C^+ \rightleftharpoons H^+ + (CH_3)_2C=CH_2$$

Finding such acids (called "superacids") turned out to be the key to obtaining stable, long-lived alkyl cations and, in general, carbocations.

If any deprotonation were still to take place, the formed alkyl cation (a strong Lewis acid) would immediately react with the formed olefin (a good π-base), leading to the mentioned complex reactions.

In Friedel–Crafts chemistry it was known that when pivaloyl chloride is reacted with aromatics in the presence of aluminum chloride, *tert*-butylated products are obtained in addition to the expected ketones. These were assumed to be formed by decarbonylation of the intermediate pivaloyl complex or cation. In the late 1950s, I returned to my earlier investigations of Friedel–Crafts complexes and extended them by using IR and NMR spectroscopy. I studied isolable complexes of acyl fluoride with Lewis acid fluorides, including higher valence Lewis acid fluorides such as SbF_5, AsF_5, and PF_5. In the course of these studies, it was not entirely unexpected that the generated $(CH_3)_3CCOF–SbF_5$ complex showed a substantial tendency toward decarbonylation. What was exciting, however, was that it was possible to follow this process by NMR spectroscopy and to observe what turned out to be the first stable, long-lived alkyl cation salt, namely, *tert*-butyl hexafluoroantimonate.

$$(CH_3)_3CCOF + SbF_5 \longrightarrow (CH_3)_3CCO^+SbF_6^- \xrightarrow{-CO} (CH_3)_3C^+SbF_6^-$$

This breakthrough was first reported in 1962 and was followed by further studies that led to methods allowing preparation of varied long-lived carbocations in solution.

The idea that ionization of alkyl fluorides to stable alkyl cations could be possible with an excess of strong Lewis acid fluoride that also serves as solvent first came to me in the early 1950s while I was still working in Hungary and studying the boron trifluoride-catalyzed alkylation of aromatics with alkyl fluorides. In the course of these studies I attempted to isolate $RF:BF_3$ complexes. Realizing the difficulty of finding suitable solvents that would allow ionization but at the same would not react with developing, potentially highly reactive alkyl cations, I condensed alkyl fluorides with neat boron trifluoride at low temperatures. I had, however, no access to any modern spectrometers to study the complexes formed. I remember a visit at the time by Costin Nenitzescu (an outstanding but not always fully recognized Rumanian chemist, who carried out much pioneering research on acid-catalyzed reactions). We commiserated on our lack of access to even an IR spectrometer. (Nen-itzescu later recalled sending his cyclobutadiene–Ag$^+$ complex on the Orient Express to a colleague in Vienna for IR studies, but the complex decomposed en route.) All I could do at the time with my RF–BF$_3$ complexes was to measure their conductivity. The results showed that methyl fluoride and ethyl fluoride complexes gave only low conductivity, whereas the isopropyl fluoride and *tert*-butyl fluoride complexes were highly conducting. The latter systems, however, also showed some polymerization (from deprotonation to the corresponding olefins). The conductivity data thus must have been to some degree affected by acid formation.

$$R-F + BF_3 \rightleftharpoons R\overset{\delta+}{-}F\overset{\delta-}{\rightarrow}BF_3 \rightleftharpoons R^+BF_4^-$$

During a prolonged, comprehensive study some years later at the Dow laboratory of numerous other Lewis acid halides, I finally hit on antimony pentafluoride. It turned out to be an extremely strong Lewis acid and, for the first time, enabled the ionization of alkyl fluorides to stable, long-lived alkyl cations. Neat SbF$_5$ solutions of alkyl fluorides are viscous, but diluted with liquid sulfur dioxide the solutions could be cooled and studied at −78 °C. Subsequently, I also introduced even lower nucleophilicity solvents such as SO$_2$ClF or SO$_2$F$_2$, which allowed studies at even lower temperatures. Following up the observation of the decarbonylation of the pivaloyl cation that gave the first spectral evidence for the tertiary butyl cation, *tert*-butyl fluoride was ionized in excess antimony pentafluoride. The solution of the *tert*-butyl cation turned out to be remarkably stable, allowing chemical and spectroscopic studies alike.

In the late 1950s, the research director of our laboratory was not yet convinced of the usefulness of NMR spectroscopy. Consequently, we had no such instrumentation of our own. Fortunately, the Dow laboratories in Midland just hundred miles away had excellent facilities run by E. B. Baker, a pioneer of NMR spectroscopy, who offered his help. To probe whether our SbF$_5$ solution of alkyl fluorides indeed contained alkyl cations, we routinely drove hundred miles in the early morning to Midland with our cooled samples and watched Ned

Baker obtain their NMR spectra. *tert*-Butyl fluoride itself showed a characteristic doublet in its 1H NMR spectrum due to the fluorine–hydrogen coupling ($J_{H,F}$ = 20 Hz). In SbF$_5$ solution, the doublet disappeared and the methyl protons became significantly deshielded from about δ1.5 to δ4.3. This was very encouraging but not conclusive to prove the presence of the *tert*-butyl cation. If one assumes that with SbF$_5$ *tert*-butyl fluoride forms only a polarized donor–acceptor complex, which undergoes fast fluorine exchange (on the NMR time scale), the fluorine–hydrogen coupling would be "washed out," while a significant deshielding of the methyl protons would still be expected. The differentiation of a rapidly exchanging polarized donor–acceptor complex from the long sought-after ionic t-C$_4$H$_9$$^+SbF_6$$^-$ thus became a major challenge.

$$(CH_3)_3C-F + SbF_5 \rightleftharpoons (CH_3)_3C \cdots F-SbF_5 \quad \text{or} \quad (CH_3)_3C^+ \ SbF_6^-$$

Ned Baker, himself a physicist, showed great interest in our chemical problem. To solve it, he devised a means to obtain the carbon-13 NMR spectra of our dilute solutions, an extremely difficult task at the time before the advent of Fourier transform NMR techniques. Labeling with carbon-13 was possible at the time only to about a 50% level (from Ba^{13}CO$_3$). When we prepared 50% ^{13}C-labeled *tert*-butyl fluoride, we could, however, obtain at best only a 5% solution in SbF$_5$. Thus the ^{13}C content in the solution was 10-fold diluted. However, Baker, undaunted, devised what became known as the INDOR (internuclear double resonance) method. Using the high sensitivity of the proton signal, he was able by the double resonance technique to observe the ^{13}C shifts of our dilute solutions—a remarkable achievement around 1960. The carbon-13 shift of the tertiary carbon atom in (CH$_3$)$_3$CF–SbF$_5$ of δ^{13}C 335.2 turned out to be more than 300 ppm deshielded from that of the covalent starting material. Such very large chemical deshielding (the most deshielded ^{13}C signal at the time) could not be reconciled with a donor–acceptor complex. It indicated rehybridization from sp3 to sp2 and at the same time showed the effect of significant positive charge on the carbocationic carbon center.

Besides the *tert*-butyl cation, we also succeeded in preparing and studying the related isopropyl and the *tert*-amyl cations.

$$(CH_3)_2CHF + SbF_5 \rightarrow (CH_3)_2CH^+ \, SbF_6^-$$

$$(CH_3)_2CFCH_2CH_3 + SbF_5 \rightarrow (CH_3)_2\overset{+}{C}CH_2CH_3 \, SbF_6^-$$

The isopropyl cation was of particular relevance. Whereas in the *tert*-butyl cation the methyl protons are attached to the carbons that are only adjacent to the carbocationic center, in the isopropyl cation a proton is directly attached to the center. When we obtained the proton NMR spectrum of the i-C_3H_7F–SbF_5 system, the CH proton showed up as an extremely deshielded septet at 13.5 ppm, ruling out a polarized donor–acceptor complex and indicating the formation of the $(CH_3)_2CH^+$ ion. The ^{13}C NMR spectrum was also conclusive, showing a very highly deshielded (by $\Delta\delta > 300$) ^+C atom ($\delta^{13}C$ 320.6). The spectrum of the *tert*-amyl cation showed an additional interesting feature due to the strong long-range H–H coupling of the methyl protons adjacent to the carbocationic center with the methylene protons. If only the donor–acceptor complex were involved, such long-range coupling through an sp3 carbon would be small (1–2 Hz). Instead, the observed significant coupling ($J_{H-H} = 10$ Hz) indicated that the species studied indeed had an sp2 center through which the long-range H–H coupling became effective. Figure 6.1 reproduces our ^1H NMR spectra of the *tert*-butyl, *tert*-amyl, and isopropyl cations. These original spectra are framed and hanged in my office as a memento, as are the ESCA spectra of the *tert*-butyl and of the norbornyl cation (vide infra).

Our studies also included IR spectroscopic investigation of the observed ions (Figure 6.2). John Evans, who was at the time a spectroscopist at the Midland Dow laboratories, offered his cooperation and was able to obtain and analyze the vibrational spectra of our alkyl cations. It is rewarding that, some 30 years later, FT-IR spectra obtained by Denis Sunko and his colleagues in Zagreb with low-temperature matrix deposition techniques and Schleyer's calculations of the spectra

Figure 6.1. ^{1}H NMR spectra of: (a) the *tert*-butyl cation [trimethylcarbenium ion, $(CH_3)_3C^+$]; (b) the *tert*-amyl cation [dimethylethylcarbenium ion, $(CH_3)_3C^+-C_2H_5$]; (c) the isopropyl cation [dimethylcarbenium ion, $(CH_3)_2C^+H$]. (60 MHz, in SbF_5:SO_2ClF solution, $-60°C$).

and varied spectroscopic methods showed good agreement with our early work, considering that our work was carried out in neat SbF_5 at room temperature long before the advent of the Fourier transform methods. Subsequently, in 1968–1970 with Jack DeMember and Auguste Commeyras in Cleveland, we were able to carry out more detailed IR and laser Raman spectroscopic studies of alkyl cations. Comparison of the data of unlabeled and deuterated *tert*-butyl cations with those of

Figure 6.2. IR spectra of *tert*-butyl (top), isopropyl (center), and *tert*-amyl (bottom) cations. T = transmission.

isoelectronic trimethylboron proved the planar structure of the carbocation.

This was also an early example of the realization that for nearly all carbocations there exists a neutral isoelectronic isostructural boron analog, which later proved itself so useful in the hands of my colleagues R. E. Williams, G. K. S. Prakash, and L. Field.

In the summer of 1962, I was able for the first time to present my work in public at the Brookhaven Organic Reaction Mechanism Conference, and subsequently in a number of other presentations and publications, I had convincing evidence to substantiate that, after a long and frequently frustrating search, persistent, alkyl cations had finally been obtained in superacidic solutions.

The chemistry of stable, long-lived (or persistent) carbocations as they became known thus began and its progress was fast and widespread. From an industrial laboratory, publication of research is not always easy. I would therefore like to thank again the Dow Chemical Company for allowing me not only to carry out the work but eventually also to publish the results.

Many have contributed since to the study of long-lived carbocations. The field rapidly expanded and allowed successful study of practically any carbocationic system. My talented and hardworking former associates and students deserve the lion's share of credit for our work, as do the many researchers around the world who joined in and contributed so much to the development of the field. (Their work can be found in the recommended readings.) I would like, however, to mention particularly the fundamental work of D. M. Brouwer and H. Hogeveen, as well as their colleagues at the Shell Laboratories in Amsterdam in the 1960s and 1970s. They contributed fundamentally to the study of long-lived carbocations and related superacidic hydrocarbon chemistry. The first publication from the Shell laboratories on alkyl cations appeared in 1964, following closely my initial reports of 1962–1963.

In the spring of 1964, I transferred from Sarnia to Dow's Eastern Research Laboratories in Framingham, Massachusetts, outside Boston, established under the directorship of a friend, Fred McLafferty, and located initially in a converted old industrial building. When Christopher Ingold came to visit us there one day later in the year, he had difficulty in convincing the cab driver who drove him from his hotel in Boston to find the "noted research laboratory" in the dilapidated neighborhood the address indicated. The laboratory, however, soon moved to neighboring Wayland, into a nice campus-like setting. Fred built up an impressive laboratory in a short time, where it was possible to work under the most pleasant conditions. Harvard and MIT were just a half-an-hour driving distance away and provided valuable contact with the academic community. Bill Lipscomb (Nobel Prize in chemistry, 1976) and Paul Bartlett from Harvard became consultants and visited regularly. The Harvard and MIT seminars provided extraordinary opportunities for participation. In 1965, however, Fred McLafferty decided to leave for academic life (first at Purdue and then Cornell). Shortly thereafter, I also felt it was time to move back to academia and followed his example.

·7·

Return to Academia— The Cleveland Years:
Carbocations, Magic Acid, and Superacid Chemistry

In Canada, following my escape from Hungary in 1956, I found no opportunity to continue my academic career. Around 1960, I had a brief visiting professorship at the University of Toronto, to which previously Alexander Todd recommended me through Charles Best. Some interest was expressed in me, but nothing developed. Years later, quite unexpectedly, I received a letter from George Wright, a senior organic faculty member at Toronto, expressing his regret that he had strongly opposed the appointment of a young Hungarian refugee chemist with no proper credentials and school ties, who he believed did not have much to offer. Belatedly, he wanted to tell me that he had been wrong. I appreciated his letter, but, as with many other things in life, one never knows what will eventually turn out to be beneficial for the future. The failure to obtain an academic appointment certainly worked out well this way in my case. I still have friends and even some family in Canada (a son of my late cousin lives in Toronto). Our younger son, Ron, was born there in 1959, and we have only pleasant memories of our Canadian years. After I received my Nobel Prize I was even rediscovered in Canada;

A Life of Magic Chemistry: Autobiographical Reflections Including Post-Nobel Prize Years and the Methanol Economy, Second Updated Edition. George A. Olah with Thomas Mathew.
© 2015 John Wiley & Sons, Inc. Published 2015 by John Wiley & Sons, Inc.

for example, the Royal Society of Canada elected me as a Foreign Fellow, which I appreciate very much.

During my stay at the Dow Eastern Research Laboratory in Boston, I regularly attended and participated in the weekly Bartlett–Westheimer seminars at Harvard. I must have made some impression, because early in 1965 Paul Bartlett asked me whether I would be interested in going back to academia. He was regularly consulting in the Cleveland–Akron area and had been asked to recommend someone as professor and chairman for the chemistry department at Western Reserve University in Cleveland. This was probably not the kind of appointment most of his former students or acquaintances were interested in. I appreciated his help, however, and was not discouraged by moving to a small Midwestern university, not well known for chemistry. Western Reserve, as I learned, was the vast reserve land that Connecticut was given in colonial times in case its crops were burned down by the Indians or ruined by some natural calamity. It was considered at the time to be on the edge of civilization but later became part of the heartland of America.

In the summer of 1965, we moved to Cleveland and I took on my duties including the chairmanship of the chemistry department. Upon my arrival in Cleveland I got much collegial help and advice from Professor Frank Hovorka, who had just retired as chair of the chemistry department and from Ernest Yeager, a leading electrochemist and faculty colleague, who acted as interim chair, and from my other new colleagues in general. I hope that I lived up to their expectations. Throughout my life I have always put great importance on loyalty, which should work both ways. It has always guided me in my relationships with colleagues and students alike, and my experience in Cleveland reinforced its importance for me.

During my chairmanship in Cleveland I was able to attract some fine colleagues to join our efforts to build a much-improved chemistry department. Among others, Miklos Bodanszky, an old friend from Budapest and an outstanding peptide chemist, came from Squibb in Princeton. With his wife Agnes (an established researcher in her own right), he soon established an excellent and most active research group in peptide synthesis. Gilles Klopman, whom I met first during a visit to the Cyanamid research laboratory in Geneva, joined us via a stay with Michael Dewar in Austin, Texas. He started one of the pioneering efforts in computational chemistry. Rob Dunbar came from John Baldeschwieler's group at Stanford and continued his work in ion cyclotron resonance spectroscopy. Muttaya Sundaralingam (who later tragically died in a Tsunami in Indonesia) came and established a fine laboratory in structural X-ray crystallography. We also attracted Bill Stephenson from Stanford, a dynamic physical organic chemist, who after a very productive time at Case Western Reserve, joined us in Los Angeles and helped to build the Loker Institute, before deciding to follow other directions in his career. George Mateescu, who originally came to do research with me from Costin Nenitzescu's Institute in Bucharest, decided to stay in America and eventually also joined the faculty. In addition to his own research, he built an outstanding instrument facility of the time centered around magnetic resonance spectroscopy, mass spectrometry, and X-ray photoelectron spectroscopy, and started close cooperation with industry.

The prime achievement of my Cleveland years of which I am particularly proud was to have succeeded in combining the chemistry departments of Western Reserve University and that of neighboring Case Institute of Technology. The two departments occupied practically adjacent buildings separated only by a parking lot. I had colleagues and good friends at Case including John Fackler and Jay Kochi. It became obvious in the year after my arrival that it would make sense to join the two departments into a single, stronger one. We achieved this by 1967 with surprisingly little friction, and I was asked to stay on as the chair of the joint department. I had however never had administrative ambitions, and I was anxious to give up as soon as feasible. In 1969 I was able to give up the chairmanship and was named the C. F. Maybery distinguished professor of research. John Fackler took over as chairman, and he did an outstanding job. All the time I continued my research uninterrupted and never gave up my teaching. Neither suffered during this period, which in fact was probably one of my most productive research periods. The merging of the two chemistry departments was so successful that it prompted the merger of Western Reserve University with Case Institute of Technology in the early 1970s, forming Case Western Reserve University. It continues as a respected research university with a particularly strong medical center in cooperation with the Cleveland Clinic.

I may have some ability for coping with administrative duties and dealing with people, but it never interfered in the ways with my research and teaching as I sometimes observed it did to others. When in my early years in Hungary I was helping to start a small research institute of the Academy of Sciences, I considered my role mainly as trying to help my colleagues while jealously safeguarding most of my time for research and scholarship. I proceeded in the same way in Cleveland and later in Los Angeles, when starting the Hydrocarbon Research Institute at USC. I have been fortunate that I never was bitten by the bug that makes many people feel important by exercising "power."

During my time as the department chair in Cleveland I discovered that my colleagues mainly came to see me with their problems. I soon realized, however, that although good things must surely have happened to them, they usually felt it not worthwhile to mention it in comparison with problems. In any case, I was able to establish a good rapport with my colleagues. They understood that I was trying to do my best to help to solve their problems within the limits of what I was able to do. This put us increasingly on the "same side." They even started to share with me some of their successes besides their problems. My efforts were not always successful, but at least they knew that I always tried my best.

Regrettably, such relationships are frequently missing in the academic setting. The ones wielding administrative "power" frequently feel obliged to mostly say "no" to whatever request is made, instead of trying to find a solution. Assumed "power" can have a strange effect on people. In my career, I have dealt with various research directors and vice presidents in industry and with department chairs, deans, provosts, and even presidents in academia, and I have learned that it is always dangerous to generalize. Some were outstanding people doing a fine job. Others, however, as a friend once characterized it, were just "swelling" in their assumed importance instead of "growing" in their jobs. There is unfortunately much infighting even in academia, perhaps—as somebody once said—because the stakes are so small. In any case, you learn much about human nature over the years, and to live with it. In academic life, leadership must be preferably by example. I am convinced that nothing is as effective as a collegial spirit of cooperation. It always was and still is a privilege for me to try to help my colleagues and students in any way I can. I also believe that some of the best academic administrators are those who are reluctant to give up their scholarship and teaching and are not only able to return full time to it but actively look forward to it (and really mean it). On the other hand, I also learned that many fine people make a deliberate choice to pursue an administrative career and are performing an essential role in leading our academic institutions and universities.

Another truth I learned about coping with administrative duties was that you can delegate much to capable and willing colleagues as well as to a small but capable staff (which you should not let grow excessively). In my experience the *de facto* time needed for making administrative decisions myself has never interfered with my regular faculty and scholarly work. This is probably the case on all levels of academia. Of course, there is also the temptation for many to spend much time in never-ending meetings filled with much self-praise. I believe that most of this is really quite unproductive and unnecessary. I always tried to be a "good citizen," never refusing committee assignments. By freely speaking my mind, however, even at the first meeting, I discovered a nearly foolproof pattern. Those running the committees almost never asked me back again and were glad to excuse me from their established "circles." I do not know whether my own experience is relevant to the general experience, but it certainly worked well for me. All this helped to preserve my ability to stay a teacher and researcher and kept me away from much other involvement. I remained for the faculty "one of us" instead of being looked upon as "one of them."

Although I may be too optimistic, I hope that our universities will regain their original goal and spirit and not succumb further to bureaucratic pressures. Universities are there for the sake of the students and the faculty to teach them. It is also their mandate to broaden our knowledge and understanding through scholarly work. Administrators at all levels should always remember that they are there to help and facilitate the academic learning and research. They should try to find solutions in a collegial spirit and not "rule" in a bureaucratic way. At the same time, they must also be capable of being decisive and making decisions even when they are not popular at the time. Am I serious about all this? Of course I am, and so are most of my colleagues whose efforts in teaching and research make our profession an honorable and essential one in an increasingly materialistic world. Regrettably, the present trend does not support my idealistic views.

The research group that followed me to Cleveland included besides my graduate student Surya Prakash (*vide infra*) two postdoctoral fellows who worked with me first ever at Dow Chris Cupas and Chuck Pittman, as well as Mel Comisarow, an extremely bright, talented young chemist who had been with me since my Sarnia days. I "rescued" him at the time when our research director declared him "difficult to get along with." Mel, a Canadian from Northern Alberta, became my first graduate student in Cleveland and went on to a successful academic career after doing his postdoctoral work with John Baldeschwieler at Stanford. At the University of British Columbia in Vancouver he codeveloped with Jim Marshall Fourier transform ion cyclotron resonance spectrometry (FT-ICR), a powerful and most useful method.

When I moved to Cleveland, Dow donated the by then quite extensive collection of laboratory samples and chemicals, as well as other specialized equipment. We packed them into a rent-a-truck and Chris Cupas, a native Bostonian, volunteered to drive it to Cleveland. Things turned out however somewhat more eventful than expected. Chris jammed the truck into an underpass of the Memorial Drive on the Cambridge side of the Charles River, disregarding the height signs and causing a major traffic jam and emergency. Fortunately, Chris was "well connected" with the Boston Police Department, and after being extricated he eventually made it to Cleveland safely with our chemicals and equipment.

The transition back to academic life was surprisingly easy for me. My group grew and rapidly reached around 15, about equally divided between graduate students and postdoctoral fellows. I attracted interested and enthusiastic graduate students as well as fine postdoctoral fellows. My group stayed at this level, with some fluctuation, over my subsequent academic career (since I am not taking on new students anymore, it became small and is composed only of postdoctoral fellows now). I have always felt that to be an effective research advisor and mentor I must have close and regular personal contact with all of my young associates. For me, it would have been impossible to do this with a larger group. Besides the pleasure of teaching again and advising graduate students, I started to receive a steady flow of postdoctoral fellows and visiting scientists from all corners of the world. Dick Chambers from Durham was one of my earliest visitors, spending a year with us. We became good friends and still maintain contact.

My postdoctoral fellows in Cleveland over the years were a truly international group. Major universities rightly attract their students and researchers to a significant degree by their reputation. Nobody would question the value in someone's resume to have studied or researched at Harvard, Stanford, Berkeley, or Caltech. Going to a little-known Midwestern university in Cleveland, however, carried no such benefits. Therefore, I particularly appreciated the personal interest and devotion of those who joined my group. Some of my American postdoctoral fellows in Cleveland included beside others Marty Bollinger, Jack DeMember, Earl Melby, Eli Namanworth, Dany O'Brien, Tom Kiovsky, Charles Juell, Richard Schlosberg, Paul Clifford, Louise Riemenschneider, David Forsyth. Others came from many other countries of the world to my laboratory to stay for a year or two of research. Among others from Great Britain came Ken Dunne, Dave Parker, and Tony White; from Germany, Joe Lukas and Herbert Mayr; from France, Jean Sommer, August Commeyras, Alain Germain, and Jean-Marc Denis; from Switzerland, Paul Kreinenbuhl and Michael Calin; from Austria, Peter Schilling; from Italy, Giuseppe Messina and Franco Pelizza; from Spain, Greg Asensio; from Japan, Masashi Tashiro, Shiro Kobayashi, Masatomo Nojima, Jun Nishimura, Yorinobu Yamada, Norihiko Yoneda, Iwao Hashimoto, Tohru Sakakibara, and Toshiyuki Oyama; from Australia, Phil Westerman, David Kelly, Bob Spear, and Subhash Narang; from Israel, Yuval Halpern, Josef Kaspi, and David Meidar; from Taiwan, Johnny Lin; from Hong Kong, Tse-lok Ho; from Rumania, George Mateescu; and from Hungary, George Sipos besides others. These associates, together with my graduate students, formed a part of my "scientific family" which now counts some 250 overall.

The relationship of a professor and his or her graduate students is a very special one. One of the most enjoyable aspects of my life was being able to guide and help the development of my graduate students and seeing them succeed. They included, besides Mel Comisarow (who followed me from Dow in Canada through Boston to Cleveland, where he became my first graduate student), such talented and highly motivated young people as Mark Bruce, Dan Donovan, James Grant, Balaram Gupta, Alice Ku, Gao Liang, Henry Lin, Christine Lui, Chuck McFarland, Ripu Malhotra, Y. K. Mo, Dick Porter, Surya Prakash (who later in Los Angeles became my close colleague and a wonderful friend, see Chapter 8), George Salem, Jacob Shen, John Staral, Jim Svoboda, Paul Szilagyi, John Welch and others.

The major credit to holding our group together at all levels goes to my wife Judy, who as an adjunct professor filled a key role over the years. It is frequently said that the scientific career of professors is made by their students and associates who *de facto* carry out their joint research. I am no exception. However, I first had to create my own little research "enclave" with an atmosphere and spirit conducive to carrying out our work. I also feel that I was able to motivate my students and coworkers, to bring out from them talents and efforts that sometimes surprised even them. It was a most rewarding experience to see that my students, when they became interested and motivated, achieved much and turned themselves into excellent, productive, and increasingly independent researchers.

The future successful careers of my graduate students and postdoctoral fellows were pursued to a larger degree in industry than in academia although worldwide many of my students achieved the leading positions. Our American institutions preferred (and still do) pedigrees from leading universities, whereas industry looked primarily for capable and adaptable chemists with the ability to fit in and achieve in the practical world. I know that my own departments at Case Western and later at USC would rarely consider a new faculty member coming from a "modest" background. It may not only be snobbery but also a conviction that good people must have been associated with the leading and best universities. It is, however, my belief (as someone who never benefited from association with a leading university) that you cannot always judge someone based only on "provenance." Individual ability and drive, as well as personality and stamina to stay the course, particularly when lacking access to outstanding students or facilities, also count much. It certainly must be, however, a great advantage to be associated with leading institutions, which rather automatically provide the most favorable conditions and also the reputation facilitating someone's career. On balance, there is justification for this preference for a "pedigree," and our leading universities and their faculties certainly deserve their standing. If I were a young person starting out today and had the choice to join one of the leading chemistry departments (as a student, postdoc, or a young faculty member), I probably would do so myself. I do not regret, however, having gone the harder way, although the decision was made for me mostly by circumstances beyond my control. I always tried, however, to make the best of the available circumstances and even benefited from it.

One of the pleasures of being a university professor is the privilege of teaching. I always enjoyed it, and I still do. Over my career I have mainly taught different aspects of organic chemistry, both at the undergraduate and, later, increasingly at the graduate level. Because my own research over the years involved synthetic, mechanistic, and structural aspects of chemistry, teaching varied topics always came naturally. I hope that I was also able to convey to my students my personal experiences and a historical perspective of chemistry. I was fortunate to have actively participated in it and knew many of the key contributors of the chemistry of the second half of the twentieth century. This helped to establish a lively and somewhat personal atmosphere in my classes, which textbooks or internet-assisted self-studies alone cannot provide. I also have always believed that it is essential to "reach" my audiences and to establish a close contact, whether lecturing to students or talking at seminars, meetings, and congresses. I never used strictly prepared lectures or texts—only occasional notes to remind me what topics to cover. This forced me to prepare mentally for my lectures. By necessity, it also ensured that my lectures always varied, even if given on the same topic. You challenge yourself to give your best, and at the same time you are also the best judge of how well you succeeded. I can see no reason why students should attend lectures by a professor who only recites a textbook or available lecture notes and does not convey new aspects to the students or challenge them to participate, something that a printed text alone or even video or internet presentation cannot do.

The real value of direct interaction between a lecturer and his audience is in the interest he can arouse by presenting the topic in a challenging and perhaps personal way as well as in the give-and-take interaction of direct exchange and participation. I have always tried in my lectures to give my students not only a presentation of facts, concepts, etc., but, based on my own experience and involvement, to convey to them the fascination of exploring the intriguing world of chemistry, making it a joyful experience. I have also tried to inspire them to explore the topics discussed further on their own. This is the way students eventually grow into competent chemists, or at least the ones who gained a reasonable understanding and appreciation of chemistry. I also hope that my students were never bored by my lectures. However intriguing the new ways of electronic communication are, even online courses on the web cannot fully replace the direct student–teacher interaction and its truly interactive give and take. This unique experience, I believe, will remain a fundamental part of the educational experience. I also believe however that as useful electronically accessible systems, printed books, and periodicals at least in our libraries will remain essential for study and research alike. Reading about any topic inevitably can lead to many other aspects to be considered that are not a priori obvious or even anticipated. Just browsing the appropriate sections of a library when we do not know exactly what we are looking for can be a stimulating and irreplaceable experience.

During my Cleveland years I also initiated an active seminar program with the participation of leading chemists who came to visit us. This allowed us to create a lively and stimulating atmosphere in the department, which benefited students and faculty alike. In 1969 I organized the first of many subsequent international research symposia. The symposium was on carbocation chemistry and was attended by many of the major investigators in the field (Nenitzescu, Brown, Winstein, Dewar, Schleyer, Gillespie, Saunders, and others). When I moved to Los Angeles, these symposia became regular annual events. My group in Cleveland held its regular weekly meetings and research seminars, which remained a permanent feature over the years. On the basis of our mutual interest, Ned Arnett and John Pople (Nobel Prize in chemistry, 1998) and their research groups from Pittsburgh joined us at regular intervals, and we had joint monthly meetings alternating between Cleveland and Pittsburgh.

My research during the Cleveland years centered on continued and extended study of carbocations in varied superacidic systems as well as exploration of the broader chemistry of superacids, involving varied systems and reagents. I had made the discovery of how to prepare and study long-lived cations of hydrocarbons while working for Dow in 1959–1960. After my return to academic life in Cleveland, a main aspect of my research was directed to the exploration of the chemistry of these persistent cations of carbon compounds (carbocations). A fascinating new area of chemistry opened up by superacids. Particular interest was directed to these systems as a great variety of carbocations was found to be readily generated and studied in these enormously strong acid systems.

Over a decade of research, we were able to show that practically all conceivable carbocations could be prepared under what became known as "stable ion conditions," using various very strong acid systems (superacids) and low nucleophilicity solvents (SO_2, SO_2ClF, SO_2F_2, etc.). A variety of precursors could be used under appropriate conditions, as shown, for example, in the case of the preparation of the methylcyclopentyl cation.

^{a}FSO$_3$H-SbF$_5$ ^{b}SbF$_5$/SO$_2$ ^{c}HF/SbF$_5$/SO$_2$

A wide variety of carbocations and carbodications, including those of aromatically stabilized nature as well as stabilized by heteroatoms, were reported in the nearly 200 publications on the topic during my Cleveland years.

During this time I suggested (in 1972) naming the cations of carbon compounds "carbocations" (because the corresponding anions for long were named "carbanions"). The name stuck and was later officially adopted by the International Union of Pure and Applied Chemistry (IUPAC) for general use.

Much interest was centered on and effort was put into studying the question of whether certain carbocations represent rapidly equilibrating or static (bridged, delocalized) systems (more about this in Chapter 9).

In course of my studies, it became increasingly clear that a variety of highly acidic systems besides the originally used antimony pentafluoride systems are capable of generating long-lived, stable persistent carbocations. The work was thus extended to a variety of other superacids. Protic superacids such as FSO_3H (fluorosulfuric acid) and CF_3SO_3H

(triflic acid) as well as conjugate acids such as $HF–SbF_5$, $FSO_3H–SbF_5$ (magic acid), $CF_3SO_3H–SbF_5$, and $CF_3SO_3H–B(O_3SCF_3)_3$ were extensively used. Superacids based on Lewis acid fluorides such as AsF_5, TaF_5, and NbF_5 and other strong Lewis acids such as $B(O_3SCF_3)_3$ were also successfully introduced. The name "magic acid" for the $FSO_3H–SbF_5$ system was given by Joe Lukas, a German postdoctoral fellow working with me in Cleveland in the 1960s, who after a laboratory Christmas party put remainders of a cake candle into the acid. The candle dissolved, and the resulting solution gave a clear NMR spectrum of the *tert*-butyl cation. This understandably evoked much interest, and the acid used was named "magic." The name stuck in our laboratory. I think it was Ned Arnett who learned about it during one of his visits and subsequently introduced the name into the literature, where it became used quite generally. When a former graduate student of mine, Jim Svoboda, started with my help a small company (Cationics) to make some of our superacidic systems and reagents commercially available, he obtained trade name protection for Magic Acid, and it has been marketed as such since.

Superacids in bottles, Olah's Lifeblood

I would like to credit here especially the fundamental contributions of Ron Gillespie to strong acid (superacid) chemistry and also to recall his generous help while I was still working at the Dow Laboratories in Canada. I reestablished contact with him during this time. We first met in the winter of 1956 at University College in London, where he worked with Christopher Ingold. Subsequently, he moved to McMaster University in Hamilton, Ontario. In the late 1950s, he already had one of the early NMR spectrometers and allowed us to study SbF_5-containing highly acidic systems carbocations. We were gratified for his help. His long-standing interest in fluorosulfuric acid and our studies of SbF_5-containing systems found common ground in studies of $FSO_3H–SbF_5$ systems.

Until the late 1950s chemists generally considered mineral acids, such as sulfuric, nitric, perchloric, and hydrofluoric acids, to be the strongest acid systems in existence. This has changed considerably as extremely strong acid systems—many billions times stronger than sulfuric acid—have been discovered.

The acidity of aqueous acids is generally expressed by their pH, which is a logarithmic scale of the hydrogen ion concentration (or, more precisely, of the hydrogen ion activity). pH can be measured by the potential of a hydrogen electrode in equilibrium with a dilute acid solution or by a series of colored indicators. In highly concentrated acid solutions or with strong nonaqueous acids, the pH concept is no longer applicable, and acidity, for example, can be related to the degree of transformation of a base to its conjugate acid (keeping in mind that this will depend on the base itself). The widely used so-called Hammett acidity function Ho relates to the half protonation equilibrium of suitable weak bases. The Hammett acidity function is also a logarithmic scale on which 100% sulfuric acid has a value of Ho –11.9. The acidity of sulfuric acid can be increased by the addition of SO_3 (oleum). Ho of anhydrous HF is –11.0 (however, when HF is completely anhydrous its Ho is –15, but even a slight amount of water drops the acidity to –11, as shown by Gillespie).

Perchloric acid ($HClO_4$; Ho –13.0), fluorosulfuric acid (HSO_3F; Ho –15.1), and trifluoromethanesulfonic acid (CF_3SO_3H; Ho –14.1) are considered to be superacids, as is truly anhydrous hydrogen fluoride. Complexing with Lewis acidic metal fluorides of higher valence, such as antimony, tantalum, or niobium pentafluoride, greatly enhances the acidity of these systems.

In the 1960s Gillespie suggested calling protic acids stronger than 100% sulfuric acid "superacids." This arbitrary but most useful definition is now generally accepted. It should be mentioned, however, that the name "superacid" goes back to J. B. Conant of Harvard, who used it in 1927 in a paper in the *Journal of the American Chemical Society* to denote acids such as perchloric acid, which he found stronger than conventional mineral acids and capable of protonating such weak bases as carbonyl compounds. My book, on superacids, published in 1985 with Surya Prakash and Jean Sommer and a second updated edition with Árpád Molnár in 2009, was appropriately dedicated to the memory of Conant. Few of today's chemists are aware of his contributions to this field. Conant subsequently became the president of Harvard University and gave up chemistry, which may explain why he never followed up on his initial work on superacids. At the end of World War II, he became the first allied high commissioner of occupied Germany and helped to establish a democratic West Germany. Upon his return home he continued a distinguished career in education and public service.

In a generalized sense, acids are electron acceptors. They include both protic (Brønsted) acid and Lewis acids such as $AlCl_3$, BF_3 that have an electron-deficient central metal atom. Consequently, there is a priori no difference between Brønsted (protic) and Lewis acids. In extending the concept of superacidity to Lewis acid halides, I suggested those stronger than anhydrous aluminum chloride (the most commonly used Friedel–Crafts acid) to be considered super Lewis acids. These superacidic Lewis acids include such higher valence fluorides as antimony, arsenic, tantalum, niobium, and bismuth pentafluorides. Superacidity encompasses both very strong Brønsted and Lewis acids and their conjugate acid systems.

Friedel–Crafts Lewis acid halides form with proton donors such as H_2O, HCl, and HF conjugate acids such as $H_2O \cdot BF_3$, HCl-$AlCl_3$, and HF-BF_3, which ionize to $H_3O^+BF_3OH^-$, $H_2Cl^+AlCl_4^-$, and $H_2F^+BF_4^-$, etc. These conjugate Friedel–Crafts acids have *Ho* values greater than –15. Thus they are much stronger than the usual mineral acids. Even stronger superacid systems are HSO_3F–SbF_5 (magic acid), HF–SbF_5 (fluoroantimonic acid), or CF_3SO_3H–$B(O_3SCF_3)_3$ (triflatoboric acid). The acidity of anhydrous HF, HSO_3F, or CF_3SO_3H increases drastically upon addition of Lewis acid fluorides such as SbF_5, which form large complex fluoroanions facilitating dispersion of the negative charge.

$$2\,HF + 2\,SbF_5 \rightleftharpoons H_2F^+ Sb_2F_{11}^- \quad \text{fluoroantimonic acid}$$

$$2\,HSO_3F + 2\,SbF_5 \rightleftharpoons H_2SO_3F^+ Sb_2F_{10}(SO_3F)^- \quad \text{magic acid}$$

$$2\,CF_3SO_3H + B(O_3SCF_3)_3 \rightleftharpoons CF_3SO_3H_2^+ B(O_3SCF_3)_4^- \quad \text{triflatoboric acid}$$

The acidity function of HSO_3F increases on addition of SbF_5 from −15.1 to −23.0, the acidity of 1:1 FSO_3H–SbF_5 (magic acid). Fluoroantimonic acid is even stronger; with 4 mole % SbF_5, the Ho value for HF–SbF_5 is already −21.0, a thousand times stronger than the value for fluorosulfuric acid with the same SbF_5 concentration. The acidity of the 1:1 HF–SbF_5 system or those with even higher SbF_5 concentrations reaches Ho −28. Thus these superacidic systems can be 10^{16} times stronger than 100% sulfuric acid! (A trillion is 10^{12}).

Related superacid systems in which SbF_5 is replaced by AsF_5, TaF_5, NbF_5, etc., are of somewhat lower acidity but are still extremely strong acids. So is HF–BF_3, a very useful superacid that will not cause oxidative side reactions. Ternary superacid systems including, for example, FSO_3H–HF or CF_3SO_3H–HF with Lewis acid fluorides are also known and used.

Acids are not limited to liquid (or gaseous) systems. Solid acids also play a significant role. Acidic oxides such as silica, silica–alumina, and zeolitic substances are used extensively as solid acid catalysts. New solid acid systems that are stronger than those used conventionally are frequently called solid superacids.

As applications of liquid superacids gained importance, efforts were directed to finding solid or supported superacids suitable as catalysts. There are considerable difficulties in achieving this goal. For example, BF_3 cannot be well-supported on solids because its high volatility makes its desorption inevitable. SbF_5, TaF_5, and NbF_5 have much lower vapor pressures and are thus much more adaptable to being supported or attached to solids. Because of their high chemical reactivity, SbF_5, HF–SbF_5, HSO_3F–SbF_5, etc., are preferentially supported on fluoridated alumina or fluorinated graphite. Solid superacids based on TaF_5 or NbF_5 are more stable than those based on SbF_5 because of their higher resistance to reduction. Solid perfluorinated resin-sulfonic acid catalysts, such as those based on the acid form of DuPont's Nafion ionomer membrane resin, and some higher perfluoroalkanesulfonic acids, such as perfluorodecanesulfonic acids, have gained use as solid superacid catalysts. Some of the widely used zeolite catalysts (based on aluminosilicates, phosphates, etc.) such as HZSM-5 are also recognized to possess high acidity.

Starting in the late 1950s, I was fortunate to have found superacidic antimony pentafluoride, magic acid, and other related systems and to be able to explore their remarkable chemistry. The strength of some of these acids can be up to many billions or even trillions times stronger than that of concentrated sulfuric acid. Such large numbers have little meaning in our everyday life, and it is even difficult to comprehend their magnitude. As a comparison, the US national debt is about 17 trillion dollars (17×10^{12}). Superacids are indeed extremely strong, considering that they are obtained in the condensed state, where the "naked" proton cannot exist. In the gas phase—for example, in a mass spectrometer at high vacuum—in contrast, the proton can be unencumbered, that is, "naked." This could be estimated to add an additional 30–35 powers of tens to the imagined (or unimaginable) Ho acidity. In the condensed state, a proton lacking an electron always will attach itself to any electron donor. It is with this caveat that we use "H^+" to denote the proton when discussing its role in condensed state chemistry.

The high acidity of superacids makes them extremely effective protonating agents and catalysts. They also can activate a new chemistry of a wide variety of extremely weakly basic compounds (nucleophiles) that previously could not be considered reactive in any practical way. Superacids such as fluoroantimonic or magic acid are capable to protonate not only π-donor systems (aromatics, olefins, and acetylenes) but also what is called σ-donors, such as saturated hydrocarbons including methane (CH_4), the simplest parent saturated hydrocarbon.

Protonated methane (CH_5^+) does not violate the octet rule of carbon. A bonding electron pair (responsible for covalent bonding between C and H atoms) is forced into sharing with the proton, resulting in 2 electron–3 center bonding (2e–3c) (see Chapter 10). Higher alkanes are protonated similarly.

$$H\!:\!\overset{\displaystyle H}{\underset{\displaystyle H}{C}}\!:\!H \;+\; \text{``H}^+\text{''} \;\rightleftharpoons\; \left[H\!:\!\overset{\displaystyle H}{C}\!\overset{H}{\underset{H}{\diagup}} \right]^+$$

Dihydrogen (H_2) is similarly protonated to H_3^+ by superacids, as was shown by studies using isotopic labeling. The structure of H_3^+ again involves 2e–3c bonding.

$$H\text{-}H \;+\; \text{``H}^+\text{''} \;\rightleftharpoons\; \left[H\!\diagdown\!\overset{H}{\underset{H}{\diagup}} \right]^+$$

$$:\!\overset{..}{\underset{..}{Xe}}\!: \;+\; \text{``H}^+\text{''} \;\rightleftharpoons\; :\!\overset{..}{\underset{..}{Xe}}\!:\!H^+$$

Nonbonded electron pair donors (*n*-donors) are expectedly readily protonated (or coordinated) with superacids. Remarkably, this includes even xenon, long considered an "inert" gas. The protonation of some π-, σ- and *n*-bases and their subsequent ionization to carbocations or onium ions can be depicted as

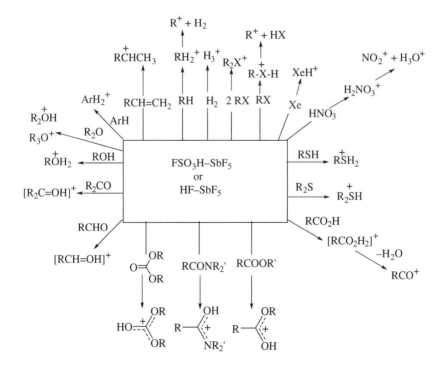

As expected, superacids were found to be extremely effective in bringing about protolytic transformations and other reactions of hydrocarbons.

Isomerization (rearrangement) of hydrocarbons is of substantial practical importance. Straight-chain alkanes obtained from petroleum oil generally have poorer combustion properties (expressed as low octane numbers of gasoline) than their branched isomers, hence the need to convert them into the higher octane branched isomers. Isomerizations are generally carried out under thermodynamically controlled conditions and lead to equilibria. As a rule, these equilibria favor increased amounts of the higher octane branched isomers at lower temperatures. Conventional acid-catalyzed isomerization of alkanes is carried out with various catalyst systems at temperatures of 150–200°C. Superacid-catalyzed reactions can be carried out at much lower temperatures (even at or below room temperature), and thus they give increased amounts of preferable branched isomers. This, together with alkylation, is of particular importance in the manufacture of lead-free high-octane gasoline. We have studied these processes extensively.

Alkylation combines lower-molecular-weight saturated and unsaturated hydrocarbons (alkanes and alkenes) to produce high-octane gasoline and other hydrocarbon products. Conventional paraffin–olefin (alkane–alkene) alkylation is an acid-catalyzed reaction, such as combining isobutylene and isobutane to isooctane.

$$(CH_3)_3CH + (CH_3)_2C=CH_2 \xrightarrow{\text{acid}} (CH_3)_2CHCH_2C(CH_3)_3$$

The key initiation step in cationic *polymerization* of alkenes is the formation of a carbocationic intermediate, which can then interact with excess monomer to start propagation. We studied in some detail the initiation of cationic polymerization under superacidic, stable ion conditions. Carbocations also play a key role, as I found not only in the acid-catalyzed polymerization of alkenes but also in the polycondensation of arenes as well as in the ring opening polymerization of cyclic ethers, sulfides, and nitrogen compounds. Superacidic oxidative condensation of alkanes can even be achieved, including that of methane, as can the co-condensation of alkanes and alkenes.

Many superacid-catalyzed reactions were found to be carried out advantageously not only using liquid superacids but also over solid superacids including Nafion-H or certain zeolites. We carried out extensive studies on the catalytic activity of Nafion-H and related solid acid catalysts (including supported perfluorooctanesulfonic acid and its higher homologues), antimony pentafluoride and tantalum pentafluoride complexed to fluorographite, etc., but we ourselves did not study zeolites.

In addition to the study of reactions of hydrocarbons and their carbocationic intermediates, in my research on superacidic chemistry, I also pursued other interests, such as the study of miscellaneous *onium ions* and derived varied reagents. Onium ions are the positively charged higher valence (higher coordination) ions of nonmetallic elements. They can be obtained by protonation (or alkylation pioneered by Meerwein) of their lower valence parents or by ionization of appropriate precursors. Many onium ions are isolable as stable salts and are also recognized as electrophilic reaction intermediates. The onium ions we studied in superacidic media or as isolated salts included acyl cations, carboxonium, carbosulfonium, carbazonium, various oxonium, sulfonium, selenonium, telluronium phophonium, halonium, siliconium, and boronium ions. I have reviewed these fields and my studies in some detail in two monographs entitled *Onium Ions* (written with colleagues) and *Halonium Ions*.

In Cleveland, I also continued my early fascination with organofluorine compounds and their chemistry. Fluorination of organic compounds requires special techniques not usually available in the average laboratory. Reactions with the most generally used and inexpensive fluorinating agent, anhydrous hydrogen fluoride, must be carried out under pressure in special equipment because of its relatively low boiling point (20°C) and corrosive nature. It is also an extremely toxic and dangerous material to work with. "Taming" of anhydrous hydrogen fluoride was thus a challenge.

We found a simple way to carry out anhydrous hydrogen fluoride reactions at atmospheric pressure in ordinary laboratory equipment (polyolefin or even glass) by using the remarkably stable complex formed between pyridine and excess hydrogen fluoride. HF (70%)

and pyridine (30%) form a liquid complex, $C_5H_5NH^+(HF)_xF^-$, showing low vapor pressure at temperatures up to 60°C. This reagent (pyridinium polyhydrogen fluoride, sometimes called Olah's reagent) thus enables us to carry out a wide variety of synthetically very useful fluorination reactions safely and under very simple experimental conditions.

We also developed a number of other useful new fluorinating reagents. They included a convenient *in situ* form of sulfur tetrafluoride in pyridinium polyhydrogen fluoride, selenium tetrafluoride, and cyanuric fluoride. We introduced uranium hexafluoride (UF_6), depleted from the U-235 isotope, which is an abundant but by law stored by-product of enrichment plants, as an effective fluorinating agent.

Studying alkylations, we developed a series of selective ionic alkylating agents. Although Meerwein's trialkyloxonium and dialkoxycarbenium salts are widely used as transfer alkylating agents, they lack selectivity and generally are incapable of C-alkylation.

In contrast, dialkylhalonium salts such as dimethylbromonium and dimethyliodonium fluoroantimonate, which we prepared from excess alkyl halides with antimony pentafluoride or fluoroantimonic acid and isolated as stable salts (the less stable chloronium salts were obtained only in solution), are very effective alkylating agents for heteroatom compounds (Nu = R_2O, R_2S, R_4N, R_4P, etc.) and for C-alkylation (arenes, alkenes).

$$2\ RX \xrightarrow[\text{or SbF}_5]{\text{HSbF}_6} \overset{+}{R}XR + SbF_6^-$$

$R = CH_3, C_2H_5$, etc. $X = I, Br, Cl$

Because the nature of the halogen atom can be varied, these salts show useful selectivity in their alkylation reactions. We also prepared varied other halonium ions and studied their alkylating ability.

$X = I, Br$

During my Cleveland years, I also continued and extended my studies in nitration, which I started in the early 1950s in Hungary. Conventional nitration of aromatic compounds uses mixed acid (mixture of nitric acid and sulfuric acid). The water formed in the reaction dilutes the acid, and spent acid disposal is becoming a serious environmental problem in industrial applications. Furthermore, because of its strong oxidizing ability, mixed acid is ill-suited to nitrate many sensitive compounds. We developed a series of efficient new nitrating agents and methods to overcome these difficulties. The use of readily prepared and isolated stable nitronium salts, such as $NO_2^+BF_4^-$, which I started to study in Hungary, was extended, and they became commercially available. These salts nitrate aromatics in organic solvents generally in close to quantitative yields, as well as a great variety of other organic compounds.

Some of the nitration reactions we studied with NO_2^+ salts were

We also found N-nitropyridinium salts such as $C_5H_5N^+NO_2BF_4^-$ as convenient transfer nitrating reagents in selective, clean reactions. Transfer nitrations are equally applicable to C- as well as to O-nitrations, allowing, for example, safe, acid-free preparation of alkyl nitrates and polynitrates from alcohols (including nitroglycerine).

To solve some of the environmental problems of mixed-acid nitration, we were able to replace sulfuric acid with solid superacid catalysts. This allowed us to develop a novel, clean, azeotropic nitration of aromatics with nitric acid over solid perfluorinated sulfonic acid catalysts (Nafion-H). The water formed is continuously azeotroped off by an excess of aromatics, thus preventing dilution of acid. Because the disposal of spent acids of nitration represents a serious environmental problem, the use of solid acid catalysts represents a significant improvement.

In Cleveland office, 1976. The pictures on the wall are of Meerwein, Ingold, Winstein, Brown, and Whitmore

All in all, my Cleveland years were most rewarding. Judy was able to rejoin me in our research as a key member of our group and I succeeded in building up a fine research group. In a relatively short time, our

In the Cleveland laboratory with Judy around 1976

efforts with my faculty colleagues resulted in creating a strong, well-balanced chemistry department in a newly emerging and dynamic university. I felt a strong attachment to my university and my colleagues, many of whom I had recruited to come to Cleveland. By the mid-1970s, however, I began to realize that what I had set out to achieve was basically accomplished. Chemistry at CWRU was by now well established, and my moving on would not severely affect it. Indeed, over the years CWRU's chemistry department has continued to be strong and dynamic, giving me great satisfaction that I had a role in shaping it.

· 8 ·

Moving to Los Angeles:
Hydrocarbon Research and Building
the Loker Institute

We enjoyed our years in Cleveland. It was also a quiet surrounding to bring up our two sons. Judy was also able to rejoin me in our research. Because our home in Shaker Heights was only about a 10-minute drive from the university, it was possible to go home for lunch, and Judy miraculously managed to keep our family going with two growing boys in the house and a husband, as usual, preoccupied with his professional life. However, after 12 years, in 1977 it was time to move on. Our younger son, Ron, was finishing high school, and our older son, George, was a senior at Case Western Reserve, majoring in accounting. Ron set his heart on going to Stanford and told us for quite a while about the wonderful life in California. He felt it would be nice for the whole Olah family to resettle there.

Our natural inclination to stay gradually weakened. Case Western Reserve University had given me a wonderful opportunity to return to academic life. I enjoyed the friendly atmosphere, my colleagues and students, and the challenge of helping to build a good chemistry department in an evolving university. However, it became increasingly clear that, despite all its positive aspects, my work in Cleveland had probably reached its limits. The excellent medical school at Case Reserve affiliated with the neighboring Cleveland Clinic represented (and continues

A Life of Magic Chemistry: Autobiographical Reflections Including Post-Nobel Prize Years and the Methanol Economy, Second Updated Edition. George A. Olah with Thomas Mathew.
© 2015 John Wiley & Sons, Inc. Published 2015 by John Wiley & Sons, Inc.

to do so) a very strong center for biomedical research. Similarly, the school of engineering and its programs in polymers and materials had gained well-deserved national recognition. My own interests and what I felt was the direction in which I wanted to extend my chemistry, that is, the broad general field of hydrocarbon chemistry, however did not seem to fit into these programs.

When I came to Cleveland as chairman of the chemistry department, Howard Schneiderman, a dynamic biologist, headed his department. Howard had foreseen the explosive development of modern biology and tried to build a significant but perhaps too ambitious biology center. He soon realized that his plans were outgrowing the possibilities offered by a moderately sized private Midwestern university. He left for the new campus of the University of California at Irvine, where as dean he built a strong biological sciences program. Subsequently he moved on to Monsanto in St. Louis, where as vice president for research he helped to transform Monsanto into a major biotechnology company and built a campus-like large (maybe too large) research complex. My own ambitions were much more modest, and I was never tempted to give up my active academic career even as I was offered some interesting administrative industrial opportunities over the years.

To move to another university was something I did not consider seriously for a long time. There was some talk about Princeton before my friend Paul Schleyer decided to move from Princeton to Germany and some about UCLA after Saul Winstein's untimely death, but nothing realized. This had also been the case, when, unbeknownst to me (I read about it only subsequently in a book written about him by one of his former students, Ken Leffer), Christopher Ingold recommended me in the mid-1960s as his successor at University College of the University of London. In the early 1970s, King's College, also of the University of London, offered me its Daniell Chair in Chemistry, but I decided to stay in Cleveland. I had no intention of leaving my adopted country, which had offered me and my family a new life and home.

In the fall of 1976 I had a call from a friend, Sid Benson, who, after a decade at the Stanford Research Institute, just returned to the University of Southern California (USC) in Los Angeles. He invited me for a visit, telling me about USC's plans to build up selected programs, including chemistry. I visited USC and found it, with its close to downtown urban campus, quite different from the sprawling expanse of the crosstown campus of UCLA, which I had visited on a number of occasions during Saul Winstein's time. I knew little about USC, except that with my family we used to watch on TV the New Year's Day Rose Bowl parade and football game (in which USC was a frequent participant). We had marveled at the sunny, summerlike weather of Southern California while we were freezing in snowbound Cleveland.

At this point, USC wanted to shed some of its football school image, or at least this was the goal of its dynamic and foresighted executive vice president Zohrab Kaprielian, who was at the time the driving force at USC. Kaprielian was an Armenian-born electrical engineer who got his education at Caltech and devoted his life to improving USC. He first built a fine electrical engineering department as its chair and followed by building up the School of Engineering as its Dean. When he became the executive vice president (the equivalent of provost) under President Jack Hubbard, he started a major effort to raise USC into the rank of major research universities. He was very persuasive in telling me about his plans but at the same time realistic about the obvious limitations. No university since Stanford's remarkable achievement in the 1950s and 1960s had been in a position to make significant improvement across the whole academic spectrum. Kaprielian thus focused on a number of areas in which he felt an impact could be made. He believed that if he could create some "centers of excellence" they would subsequently help to raise the level of the whole university. He was looking for people who he felt had promise and drive as well as a willingness to take a chance on building up something from modest beginnings.

We hit it off well. I liked his approach and his outspoken honesty about his goals. He also must have seen something in me that he liked, and we started seriously discussing my coming to USC. There were, however, difficulties. The chemistry department of USC was heavily centered around chemical physics (at the time, concentrating on spectroscopy). Organic chemistry for whatever reasons was much of a stepchild, although Jerry Berson spent a decade at USC before he left for Wisconsin and then Yale. Ivar Ugi was also a faculty member for 3 years (1969–1971) before returning to Germany. He laid the foundation of multicomponent synthesis, that is, combinatorial chemistry while at USC, although it did not attract much attention at the time. The facilities for experimental work were poor, with inadequate wet laboratory space. In our discussions, I mentioned to Kaprielian my interest in significantly extending my previous work into the area of hydrocarbon chemistry. I felt that by establishing a strong program of basic research and graduate education in hydrocarbon chemistry, USC could become a leader in this important field. Because the memory of the first Arab oil embargo was still fresh, this struck a cord with Kaprielian, who felt that he could "sell" my research interest to the trustees and establish a "Hydrocarbon Research Institute" at USC that could accommodate me, as well as other chemistry faculty members whose interests could be fitted into its framework.

I got much advice and encouragement during my deliberations over moving to USC from Martin Kamen, a friend and true Renaissance man, as well as an outstanding scientist (codiscoverer of the carbon-14 isotope and a leading biological chemist). Martin was at the time attempting to build up the molecular biology program at USC. Things never really worked out for his plans, but, characteristically, he looked at the broader picture and encouraged me. He also taught me that, this being close to Hollywood, not everything is as it seems or as it is promised, and thus it is useful to have matters properly clarified and put into writing.

USC was at the same time also attempting to attract Franz Sondheimer from the University of London, a fine organic chemist whose research interest in annulenes also fit into the broad scope of hydrocarbon chemistry. The two of us, together with Sid Benson whose physicochemical research rounded out our developing plans, seemed to form an attractive group. Benson enthusiastically supported the effort and so did Jerry Segal, who as department head eased the way and solved many problems. In December 1976, both Franz Sondheimer and I agreed to come to USC. We bought a house that same month in a canyon on top of Beverly Hills, on the same street where the Sondheimers decided to live. At my son Ron's urging, we built a swimming pool, which we use ever since for our daily morning exercise and for our enjoyment. Regrettably, Franz, who in some ways was a troubled man, changed his mind twice and eventually withdrew. Some years later, he tragically ended his life, cutting short a remarkable career that much enriched chemistry.

Whenever I make up my mind about something, I never look back and move forward determinedly. Once I decided to join USC, I was planning to move as soon as the coming spring, despite the fact that there were no adequate laboratory facilities available to accommodate my group of about 15 graduate students and postdoctoral fellows who decided to make the move west with me. Kaprielian put together a plan to build our own modest institute building of some 17,000 sq. ft., which was approved by the trustees after I had a chance in April 1977 to address them at their meeting in Ojai, a pleasant resort town not far from Los Angeles. It was promised to be ready for occupancy in 2 years, and the university committed the initial funds to finance it. In the meantime, some temporary space was found for us in the basement of what was called the "old science building" (which lived up to its name). It took some faith to accept this arrangement, particularly because I was determined that our move should not interrupt the research effort of my group. My wife Judy, who is more practical than I am, had serious doubts that the schedule would work out, but eventually it did. In December 1979, two-and-a-half years after our move, we were able to occupy the new Hydrocarbon Research Institute building.

By a mutually agreeable arrangement with Case Western, I was able to take most of my laboratory equipment, chemicals, and instrumentation with me to Los Angeles. I again got help from Dow Chemical, which also donated an additional NMR spectrometer (of great value at that time). We packed up our laboratories in Cleveland in late May, with my graduate students John Welch and Surya Prakash spearheading and organizing the effort. Our small convoy, including moving vans, started out across the country to California in the best spirit of westward migration. Most of my group drove to Los Angeles, but Judy and I decided to fly out with our cocker spaniel Jimmy in the baggage compartment. My sons drove out in our car, getting also to know the country and each other better during the long trip.

The arrival of the Olah group at the USC campus with its moving vans caused quite a stir. Whatever shortcomings our temporary quarters had were overcome by our enthusiasm, and, miraculously, in 3 weeks we were back doing research. I am not sure whether everybody at USC was pleased by this "invasion" and our determination to quickly overcome any difficulties to get started. My background may have prepared me to manage as best as possible in whatever circumstances I found myself, but it was the enthusiasm and hard work of my students that made the nearly impossible possible. I must say, however, that it was not an easy start. My group found itself isolated in many ways and left to its own resources. Organic chemists at USC were still considered somewhat outsiders. Furthermore, there is perhaps something in human nature that gives some people satisfaction to see others struggle and may be even fail. In our case, failure would have justified the assumption by some that at USC it was not possible to succeed in organic chemistry. The Germans have a word "Schadenfreude" to express such feelings (meaning the joy of misfortune of others). Probably it would have been easier to accept us if it had taken a much longer struggle for us to settle in. In any case, we persevered and succeeded.

Early in my career, someone told me that if I ever became even mildly successful, there would always be those who would envy me. I was also told, however, that this was still better than if they felt sorry for my failure or misfortune. I always remembered this advice, and it helped me to understand human nature and not to be easily offended. Universities also have their own internal politics, disagreements, and differing views. Academic infighting, as it is said, can be fierce because the stakes are so small. I always felt, however, that it is much wiser to stay out of it as much as possible and to use my energy for much more rewarding efforts in research, scholarly work, and teaching. Thus, once I realized that it would be very difficult to change the mind and views of some of my faculty colleagues, I simply went on to pursue my own work without arguing or imposing on others. I am grateful for the help and understanding I received at USC from many. However, bringing about, what I considered (and in some way still consider) necessary changes proved to be difficult. I therefore put my efforts single-mindedly toward building the Hydrocarbon Institute into a viable entity within the university with a fair degree of independence and endowment too be raised to pursue its research and educational goals. We also set our own standards.

As I mentioned, USC provided some start-up funds for the Institute. Further development and operation, however, was only possible through raising support. I never had been seriously involved in such an effort previously, but I consider myself a reasonably fast learner. During the mentioned trustee meeting at Ojai in the spring of 1977, Judy and I met with the leaders of the "Trojan family" (as the USC community is known), including many wonderful people. Carl Franklin, vice president at the time, and his wife Caroline were particularly kind to us. The Franklins became friends, and Carl became a major supporter of our efforts.

Soon after our move to Los Angeles, Carl Franklin introduced me to Katherine and Don Loker, friends of his and benefactors of USC (as well as of other institutions including Harvard University). Katherine, a native Angelina, was a graduate of USC, where she was also a star athlete. Her father had emigrated from the Adriatic coast of Croatia to California, eventually building up a large tuna fishing and canning

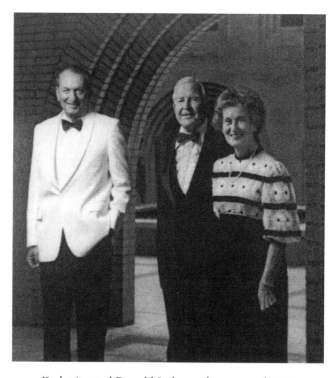

Katherine and Donald Loker and me around 1985

company (Starkist). Don was a New Englander from Boston, who after some time at Harvard moved out west and became an accomplished actor (under his professional name of Don Winslow) with some 50 movies to his credit. He later joined Starkist and became a successful businessman. Katherine and Don were married before World War II. Don's military service was in the Pacific, where he ended up on the staff of General MacArthur (a family friend) in Tokyo and helped to reestablish the Japanese fishing industry. During his stay in Tokyo he also learned Japanese quite well. I remember, some years later, when Reiko Choy became my secretary. She also was a wonderful lady of Japanese origin, who for many years kept my office operating smoothly and also managed miraculously to turn my terrible handwritten papers and manuscripts, including that of the first edition of this book, into proper form. When Don Loker came to visit me, they started to talk in Japanese (quite a surprise coming from a New England Yankee). I still write my manuscripts in long hand, to the annoyment of my fully computer-adapted wife, even refusing to use a computer, I believe that this slow and tedious process gives me more time to think over my thoughts. In any case I may miss some wonderful technical progress but it still works for me.

The Lokers had no background or business interest in chemistry. It was Carl Franklin, who told them about USC's effort to establish a

With Carl Franklin

hydrocarbon research institute, and he must have also told them something kind about me. Don's first reaction, as I understand it, was however "What the hell are hydrocarbons?" Nevertheless, they eventually became interested and in 1978 made their first gift to the Institute, which in 1983 was renamed to honor them as the "Loker Hydrocarbon Research Institute." In 1991, USC under its president Steven Sample's foresighted leadership formally adopted "Guidelines" concerning the Institute's organization and its permanence. The Lokers continued their generosity toward the Institute, which, together with that of some other friends and supporters, particularly of Harold Moulton, was invaluable to our efforts. Their friendship and sound advice, however, was equally valuable. Don Loker passed away in 1989, but Katherine continued her support and chaired the Institute's Advisory Board till her death in 2008.

My active life and work were unexpectedly affected by two serious illnesses that I fought in 1979 and 1982. We spent the first 2 years in Los Angeles in temporary laboratories in the basement of an old science building, where I also had my office on the ground floor above the laboratories. Jerry Segal occupied an adjacent office. I mention this because in early 1979 I became seriously ill with a rather mysterious disease that was difficult to diagnose but that increasingly worsened and eventually

With friends and benefactors, Harold Moulton and the Lokers

became life-threatening. It turned out to be a rare immune disease, which manifested itself in bad sores breaking out all over my body, which were not healing and reached frightening levels. I was seen by many dermatologists in Los Angeles (probably a mistake, because the problem was much deeper than a skin problem), but none of them could find the cause. Without realizing any connection at the time, Jerry Segal also started to show the same symptoms. By August, I was in such poor health that Judy decided as a last resort that we should go to the Mayo Clinic in Rochester, MN. This probably saved my life. Within 24 hours after our arrival the Mayo doctors diagnosed my problem as an acute case of pemphigus vulgaris and started to treat me with megadoses of Prednisone and other rather powerful medications. We planned to be at the Mayo Clinic only for a short stay for diagnosis, but I was hospitalized there for a month. Pemphigus is an immune disease, of which not much was known at the time, except that it can be triggered by certain chemicals such as penicillamine. As it turned out, in the 1940s and 1950s USC's old science building was used for pharmacological research using penicillin (and probably penicillamine). Some old air ducts of the building could still have been contaminated. Pemphigus affects only about one in a million in the United States; thus it is a true orphan disease receiving little interest. The random probability that two individuals having adjacent offices in the same building should come down with it simultaneously is astronomical.

In any case, I eventually recovered (and so did Jerry), but my immune system must have suffered serious damage, which demonstrated itself 3 years later, when I collapsed in my office one day and was found to be bleeding internally from a form of rare stomach cancer, which necessitated major surgery but was fortunately localized. I again recovered and have had no further stomach difficulties since. Whether weakening and knocking out my immune system to overcome the previous problems had any effect is not clear, but it could have been a factor. Despite my health problems I was able to continue my work without much interruption and the scientific productivity of my group has not suffered. As a matter of fact the early 1980s were some of our best years.

When you go through serious health problems (see also subsequent struggles) and face your mortality, your attitude changes and you learn to appreciate what is really meaningful and what is only peripheral in your life. It certainly affects your general views, but it also focuses you better in your scientific work. You gain a clearer understanding of what in your work may have a more lasting value and therefore is worth pursuing, instead of scattering your efforts.

Our institute building was designed in the fall and winter of 1977, working with the noted LA architect William Pereira and his firm. Jerry Segal and Tony Lazzaro, USC vice president in charge of facilities, were most helpful in the process. The institute was given a central location in the middle of the campus adjacent to other science and engineering buildings. It replaced some World War II-vintage barracks still used for smaller classes and storage. The design provided a functional building, well suited for chemical research but at the same time also an attractive home for faculty, students, and staff. We were working with a small budget ($1.7 million), which translated to the cost of $100/sq. ft. in determining the size of the building (that time it was possible to still build a fine building at very low cost). Obviously, much needed to be economized, but the plans allowed for future expansion. The University's initial financial commitment was subsequently supplemented by a generous gift of the Lokers that allowed the building to open its doors in late 1979.

Loker Hydrocarbon Research Institute at USC

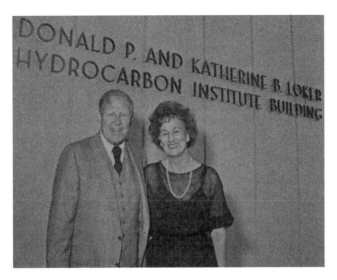

Dedication of the new institute building, 1979–Don and Katherine Loker

As the Institute's work and scope grew, so did the need for additional space. In 1989, Katherine Loker continued her tradition of generous giving with a major gift, which allowed the construction of a new wing (named the Katherine Bagdanovich Loker Wing). The new wing houses additional state-of-the-art laboratories, instrument facilities, and what became the George and Judith Olah Library and Conference Room. It contains my extensive library and periodicals collection, which we donated in 1977 when we came to USC and which we have expanded and maintained since. Additional gifts, particularly by the late Harold E. Moulton, a generous supporter and a founding member of our board, greatly advanced the Institute's development. Carl Franklin, by then emeritus vice president of USC, remained a strong supporter and was relentless till his death in pursuing support for the Institute. We are extremely grateful to them all and to others for their generosity and support. The extended Institute now has some 43,000 sq. ft. of space in a most attractive setting and provides not only first-class research facility but in a sense a home away from home for all the whole Loker Institute family, including what the younger generation named "le bistro," a pleasant setting for lunch and informal discussions.

The new addition of the building was completed by the end of 1994 and dedicated in February 1995. Because I coincidentally won the Nobel Prize just 2 months before (more about this in Chapter 11), some believed that there was some relationship between the two events. This certainly was not the case. Katherine Loker and our other friends had made their wonderful gifts well before, and it was just a fortunate coincidence that we had such "good timing" to celebrate the opening of our enlarged institute.

Physical facilities help, but do not *per se* make a research institute. It is the people who work there and their contributions and devoted hardwork that is most important. We are now well into the fourth decade since the Hydrocarbon Research Institute was started at USC. At the beginning in 1977, Sid Benson and I shared the scientific directorship of the Institute and Jerry Segal carried out the administrative responsibilities as executive director. When we moved into our own building in 1979, Bill Stephenson, physical organic chemist and a former colleague of mine in Cleveland who subsequently joined us at USC, took over as executive director. Bill, a fine scientist and teacher, was also an outstanding colleague and good administrator. In 1983, however, he decided to pursue other interests and left academia.

Katherine Loker wing of the Institute

The Olah Library

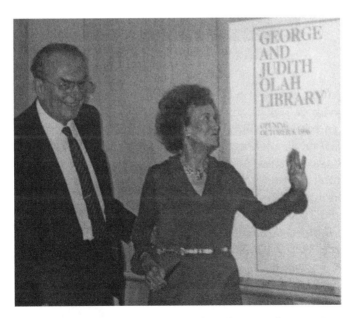

Dedication of the Olah Library with Katherine Loker in 1996

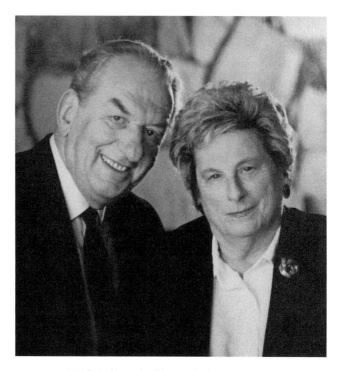

With Judy at the library dedication, 1996

To fill the void, we attracted another USC colleague John Aklonis, a polymer physicochemist, as executive director. With John joining the Institute, we made a commitment to include polymer chemistry and, subsequently, materials research as an integral effort. John became a close friend and is one of the nicest individuals I ever met. He contributed greatly to our efforts and also helped me personally to continue to stay active in my research by taking most of the burden of administrative duties off me. His outgoing personality and humanity made him loved and respected by everybody and allowed the Loker Institute to continue and expand its work. John should be given the lion's share of credit for making the expansion of our Institute a reality, helping to raise funds, planning the addition to our building, and keeping the Institute afloat in the sometimes turbulent waters of the internal politics of the university. John and his wife Jo were always avid sailors. Their life was very much centered around their sailing and they even lived on their boat. In 1990, John decided, approaching the age of 50, to give up his university career, take early retirement, and set sail to spend the next decade (or two) on the high seas. Subsequently, they returned to United States and settled in Oregon. We miss them much, but we keep in touch and they visit Los Angeles occasionally. From the Institute's and my own point of view, however, John's departure was a serious blow and he was not easily replaced. I became the overall Institute's director. Bill Weber (till his retirement) and Robert Aniszfeld, who was a graduate student of mine, became our excellent associate directors and together with our small but highly dedicated and efficient staff have taken care of most administrative duties.

In shaping the scientific work of the Loker Institute as well as to pursue my own research interests, I got invaluable help from the colleagues who joined in the effort, and particularly from G. K. Surya Prakash, my former graduate student who stayed on and became over the years a highly respected professor, my friend and partner in our joint research. He also succeeded me as director in 2010. Relieved of administrative duties, I continue my research. It is not often that outstanding scientific ability and wonderful human characteristics are combined in a person. Prakash is such a unique person. I am lucky to have been associated with him for the last four decades. I am also proud to have seen him develop into an outstanding scholar and teacher in his own right. He moved up

With Surya Prakash

in his academic career to a full professorship. Judy and I donated part of my Nobel money and established together with the generous help of Carl Franklin and other friends, a professorship at USC. To my delight, Prakash was named the first "George and Judy Olah Professor." As the director of the Loker Institute, he is doing a great job. I stay on at our President Max Nikias' request as founding director and professor of chemistry and engineering, concentrating my effort on the study of the "Methanol Economy" (see Chapter 15). At the same time, I am also chairing the advisory board of the Institute.

With Robert Aniszfeld and administrative assistant Jessie May

Other faculty colleagues in the organic/polymer area in our Institute include Thieo Hogen-Esch, Nicos Petasis, Kyung Jung, Travis Williams, Barry Thompson, and Chao Zhang. Golam Rasul, a research professor, contribute significantly to our computational chemistry efforts. From JPL, Sri. Narayan, who was a coinventor of our direct methanol fuel cell (DMFC) project joined our Institute in 2010 as a research professor and now heads the electrochemistry group. Karl Christe, an outstanding inorganic chemist and a research professor, heads an extremely productive laboratory in the chemistry of fluorine and high-energy compounds. After the untimely passing away of our friend and colleague Bob Bau, Ralf Haiges, another fine research faculty member, provides us with X-ray crystallographic studies. After Sid Benson's retirement and passing away, for a time, Marc Thompson and Larry Dalton (who was also a scientific codirector, later left USC to University of Washington) spear-headed efforts in the more physically oriented materials–catalysis area.

Since its inception, the Institute, in addition to its regular faculty members, has always had a number of outstanding colleagues associated with it as fellows or adjunct faculty. These included, among others, initially Ned Arnett, Joe Casanova, Paul Schleyer, Jean Sommer, Bob Williams, Michael Szwarc, and Ken Wade with presently great collaboration by Gabor Somorjai and Peter Stang. They all greatly contributed to our scientific work by their visits and participation in our symposia and frequently through close research cooperation. Because they were also close friends, our association added much to the enjoyment and pleasure of our efforts. I only hope that they also felt the same. Sadly, I should mention here that during the years many of my great supporters, friends, and colleagues—Carl Franklin, Don and Katherine Loker, Harold Molton, Sid Benson, Al Cotton, Ken Wade, Michael Szwarc, Paul Schleyer besides others—departed us. We cherish their memory and contributions.

With Paul Schleyer

The objectives of the Loker Hydrocarbon Research Institute have stayed the same since its creation in 1977. They are:

♦ To pursue the long-range development of hydrocarbon chemistry

♦ To develop new fuels and materials and to provide environmentally sustainable solutions to energy generation problems

♦ To train researchers in the field of hydrocarbon chemistry

♦ To further the interchange of information through publications and symposia on developments in hydrocarbon chemistry

♦ To act as an international center of hydrocarbon chemistry and to facilitate exchange of information and ideas through visits of scientists, colloquia, and research symposia

♦ To foster a close relationship with the chemical, petroleum, gas, and energy industries and governmental agencies for the exchange of information and knowledge and to ensure that research results and discoveries of significance will be effectively exploited

With Peter Stang

Hydrocarbons, as their name indicates, are compounds of carbon and hydrogen. They represent one of the most significant classes of organic compounds. The scope of hydrocarbons is broad. They include saturated hydrocarbons (alkanes, cycloalkanes) and their derivatives as well as unsaturated alkenes and dienes, acetylenes, and aromatics.

In methane (CH_4), the simplest saturated alkane, a single carbon atom is bonded to four hydrogen atoms by sharing electron pairs, that is, by covalent bonds. In the higher homologues of methane (of the general formula C_nH_{2n+2}), all atoms are bound to each other by such single [sigma (σ), two-electron-two-center] bonds, with carbon atoms also displaying their ability to bind with each other to form C–C bonds. Whereas in CH_4 the H:C ratio is 4, in C_2H_6 (ethane) it is decreased to 3, in C_3H_8 (propane) to 2.67, and so on. Alkanes can be straight chain (each carbon attached to ≤ 2 other carbon atoms) or branched (in which at least one of the carbon is attached to either three or four other carbon atoms). Carbon atoms can be aligned in open chains (acyclic hydrocarbons).

Carbon can also form multiple bonds with other carbon atoms. This results in unsaturated hydrocarbons such as olefins (alkenes), containing a carbon–carbon double bond, or acetylenes (alkynes) containing a carbon–carbon triple bond. Dienes and polyenes contain two or more unsaturated bonds.

Aromatic hydrocarbons (arenes), of which benzene is the parent, consist of a cyclic arrangement of formally unsaturated carbons, which, however, give a stabilized (in contrast to their hypothetical cyclopolyenes), delocalized system.

The H:C ratio in hydrocarbons is indicative of the hydrogen deficiency of the system. As mentioned, the highest theoretical H:C ratio possible for hydrocarbon is 4 (in CH_4), although in electron-deficient carbocationic compounds such as CH_5^+ and even CH_6^{2+} we were able to show that the ratio is further increased (to 5 and 6, respectively, see Chapter 10). On the other end of the scale in extreme cases, such as the dihydro- or methylene derivatives of C_{60} and C_{70} fullerenes, the H:C ratio can be as low as 0.03.

The broad field of hydrocarbon chemistry including our own varied work over the years was discussed in my extensive monograph (*Hydrocarbon Chemistry* by Wiley 1995, 2003) with Árpád Molnár, a professor and long-standing friend from the University of Szeged in Hungary, during his prolonged stay in our Institute and our extensive publications covering some 50 years.

Hydrocarbons are abundant in nature. All *fossil fuels* (coal, oil, gas) are basically hydrocarbons, deviating, however, significantly in their H:C ratio.

H:C Ratio of Natural Hydrocarbon Sources

Methane	4.0
Natural gas	3.8
Petroleum crude	1.8
Tar sands bitumen	1.5
Shale oil (raw)	1.5
Bituminous coal	0.8

Natural gas, a gaseous fossil fuel, is a mixture of gaseous hydrocarbons consisting primarily of methane (present usually in >80–90%), some of the higher homologous alkanes (ethane, propane, butane) and much lower percentage of carbon dioxide, nitrogen, and hydrogen sulfide. In "wet" gases the amount of C_2–C_5 alkanes is higher (gas liquids). Most natural gas was formed over time by biogenic and thermogenic mechanisms. Biogenic gas is created by methanogenic organisms in marshes, bogs, landfills, and shallow sediments while the thermogenic gas is formed when organic materials from the decayed remains from plants and animals trapped beneath the rock are exposed to intense heat and pressure over thousands of years. Therefore, natural gas is generally found in deep underground rock formations or associated with other hydrocarbon reservoirs in coal beds, shale formations, and as methane clathrates. In the offshore continental shelves and on land in arctic regions, large quantities of natural gas, primarily methane, exist in the form of hydrates or clathrates under high pressure and low temperature. It is obtained from biogas mainly landfill gas and town gas (obtained from destructive distillation of coal). Natural gas is a major energy source often used for heating, cooking, and electricity generation. It is also used as fuel for vehicles and as a chemical feedstock in the manufacture of plastics and other commercially important organic chemicals. Natural gas is a nonrenewable resource because it cannot be restored on a human time frame. Estimated world proven natural gas reserves at the beginning of 2013 was 194.9 trillion cubic metres, sufficient to meet over half a century of global production. Recently discovered way to produce shale gas adds significantly to our reserves.

Petroleum or crude oil, a liquid fossil fuel is a complex mixture of many hydrocarbons. It consists of saturated, predominantly straight chain alkanes, small amounts of slightly branched alkanes, cycloalkanes, and aromatics. Petroleum is generally believed to be derived from organic matter deposited in the sediments and sedimentary rocks on the floor of marine basins. The identification of biological markers such as petroporphyrins provides convincing evidence for the biological origin of oil (however, there may also be hydrocarbons of abiogenic origin). The effects of time, temperature, and pressure in the geological transformation of the organics to petroleum are not yet clear. However, considering the low-level of oxidized hydrocarbons and the presence of porphyrins, it can be surmised that the organics were acted upon by anaerobic microorganisms and that temperatures were moderate, <200°C. By comparing the elemental composition of typical crude oils with typical bituminous coals, it becomes clear why crude oil (as well as natural gas) is a much more suitable fuel source in terms of its higher H:C atomic ratio, generally lower sulfur and nitrogen contents, very low ash contents (probably mostly attributable to suspended mineral matter and vanadium and nickel associated with porphyrins), and essentially no water content.

It is interesting to note that recent evidence shows that even extraterrestrially formed hydrocarbons can reach the Earth. The Earth continues to receive some 40,000 tons of interplanetary dust every year. Mass spectrometric analysis has revealed the presence of hydrocarbons attached to these dust particles, including polycyclic aromatics such as phenanthrene, chrysene, pyrene, benzopyrene, and pentacene of extraterrestrial origin indicated by anomalous isotopic ratios.

Petroleum—a natural mineral oil—was referred to as early as the Old Testament. The word *petroleum* means "rock oil" (from the Greek *petros* (rock) and *elaion* (oil)). It has been found for centuries seeping out of the ground, for example, in the Los Angeles basin in what are now called the La Brea tar pits. Vast deposits were found in Europe, Asia, the Americas, and Africa.

In the United States, the first commercial petroleum deposit was discovered in 1859 near Titusville in western Pennsylvania when Edwin Drake and Billi Smith struck oil in their first shallow (~20 m) well. The well yielded some 400 gallons of oil per day (about 10 barrels). The area had already been known to contain petroleum, which residents skimmed from the surface of a local creek, called "Oil Creek." The first oil-producing well rapidly opened up a whole new industry. The discovery was not unexpected, but it provided evidence for oil deposits in the ground that could be reached by drilling. Oil was used for many purposes at that time, such as burning in lamps and even medical remedies. The newly discovered Pennsylvania petroleum was soon also marketed to degrease wool, prepare paints, fuel steam engines, power light railroad cars, and for many other uses. It was recognized that the oil was highly impure and had to be refined to separate different fractions for varied uses. The first petroleum refinery, a small stilling operation, was established in Titusville in 1860 and subsequently John D. Rockefeller started his Standard Oil empire with refinery in his hometown of Cleveland. Petroleum refining was much cheaper than producing coal oil (kerosene), and soon petroleum became the predominant source for kerosene as an illuminant. With the popularity of automobiles in the 1910s, gasoline became the major petroleum product. Large petroleum deposits were found in California, Texas, Oklahoma, and, more recently, Alaska. Areas of the Middle East, Asia, Russia, Africa, South America, and, more recently, the North Sea became major world oil production centers.

The daily consumption of crude oil in the United States is presently about 16–21 million barrels. The world consumption is about 80–90 million barrels (some 11–13 million tons) per day; the United States uses about 18–20% of the world's total but has less than 5% of the world population. Most of this oil is used for transportation fuels, varied products, and for heating. Petroleum and natural gas are used as feedstocks for the manufacturing of chemicals, pharmaceuticals, plastics, elastomers, paints, and a host of other products. Petrochemicals from hydrocarbons provide many of the necessities of modern life, to which we have become so accustomed that we do not even notice our increasing dependence, even though the consumption of petrochemicals is still growing. Advances in the petroleum–hydrocarbon industry, more than anything else, may be credited with the high standard of living we enjoyed in the late twentieth century.

Whereas light crudes are preferred in present-day refining operations, increasingly, *heavy petroleum sources* also must be processed to satisfy ever-increasing needs. These range from commercially usable heavy oil (California, Venezuela, etc.) to the huge petroleum reserves locked up in *shale* or *tar sand* formations. These more unconventional hydrocarbon accumulations exceed the quantity of lighter oil present in all the rest of the oil deposits in the world together. One of the largest accumulations is located in Alberta, Canada, in the form of large tar sand and carbonate rock deposits containing some 2.5–6 trillion barrels of extremely heavy oil called *bitumen*. There are large heavy oil accumulations in Venezuela, Siberia, among other areas. Vast, commercially significant reservoir of oil is the oil shale deposits located in Texas, Wyoming, Utah, Colorado, etc. The practical use and size of these potentially vast reserves depends on finding economical ways to extract the oil.

The quality of petroleum varies, and, according to specific gravity and viscosity, we talk about light, medium, heavy, and extra heavy crude oils. Light oils of low specific gravity and viscosity are more valuable than heavy oils with higher specific gravity and viscosity. In general, light oils are richer in saturated hydrocarbons, especially straight chain alkanes, than are heavy oils; they contain <75% straight chain alkanes and <95% total hydrocarbons. Extra heavy oils, the bitumens, have a high viscosity, and thus may be semisolids with a high heteroatom content (nitrogen, oxygen, and sulfur) and a corresponding reduced hydrocarbon content, of the order of 30–40%.

Typical compositions of light and heavy oils are given below.

Fraction	Light oil (%)	Heavy oil (%)
Saturates	78	17–21
Aromatics	18	36–38
Resins	4	26–28
Asphaltene	Trace-2	17

Heavy oils and especially bitumens contain high concentrations of resins (30–40%) and asphaltenes (<20%). Most heavy oils and bitumens are thought to be derivatives of lighter, conventional crude oils that have lost part or all of their straight chain alkane contents along with some of their low-molecular-weight cyclic hydrocarbons through processes taking place in the oil reservoirs. Heavy oils are also abundant in heteroatom (N, O, S)-containing molecules, organometallics, and colloidally dispersed clays and clay organics. The prominent metals associated with petroleum are nickel, vanadium (mainly in the form of vanadyl ions (VO^{2+})), and iron. Some of these metals are (in part) bound to porphyrins to form metalloporphyrins.

Processing heavy oils and bitumens represents a challenge for the current refinery processes, because heavy oils and bitumens poison the metal catalysts used in the refineries. In our research at the Loker Institute, we found the use of superacid catalysts, which are less sensitive to heavy oils, an attractive solution to their processing, particularly hydrocracking.

Coals (the plural is deliberately used because coal has no defined, uniform nature or structure) are solid fossil sources with low hydrogen content. The "structure" of coals means only the structural models depicting major bonding types and components relating changes with coal rank. Coal is classified, or ranked, as lignite, subbituminous, bituminous, and anthracite. This is also the order of increased aromaticity and decreased volatile matter. The H:C ratio of bituminous coal is about 0.8, whereas anthracite has H:C ratios as low as 0.2.

From a chemical, as contrasted to a geologic, viewpoint, the coal formation (coalification) process can be grossly viewed as a continuum of chemical changes, some microbiological and some thermal in origin, involving a progression in which woody or cellulosic plant materials (the products of nature's photosynthetic recycling of CO_2) in peat swamps are converted over many millions of years and under increasingly severe geologic conditions to coals. Coalification is grossly a deoxygenation–aromatization process. As the "rank" or age of the coal increases, the organic oxygen content decreases and the aromaticity (defined as the ratio of aromatic carbon to total carbon) increases. Lignites are young or "brown" coals that contain more organic oxygen functional groups than do subbituminous coals, which in turn have a higher carbon content but fewer oxygen functionalities.

The organic chemical structural types believed to be characteristic of coals include complex polycyclic aromatic ring systems with connecting bridges and varied oxygen-, sulfur-, and nitrogen-containing functionalities.

The main approaches used to convert coals to liquid hydrocarbons (coal liquefaction) center around breaking down the large, complex "structures," generally by hydrogenative cleavage reactions, and increasing the solubility of the organic portion. Coal liquefaction can be achieved by direct catalytic hydrogenation [pioneered by Bergius (Nobel Prize in chemistry, 1931)]. Combinations of alkylation, hydrogenation, and depolymerization reactions followed by extraction of the reacted coals are the major routes taken. This can provide liquid fuels, providing gasoline and heating oil.

Different types of other coal liquefaction processes have been also developed to convert coals to liquid hydrocarbon fuels. These include high-temperature solvent extraction processes in which no catalyst is added. The solvent is usually a hydroaromatic hydrogen donor, whereas molecular hydrogen is added as a secondary source of hydrogen. Similar but catalytic liquefaction processes use zinc chloride and other catalysts, usually under forceful conditions (375–425°C, 100–200 atm). In our own research, superacidic $HF–BF_3$-induced hydroliquefaction of coals, which involves depolymerization-ionic hydrogenation, was found to be highly effective at relatively modest temperatures (150–170°C).

The ultimate "depolymerization" of coal occurs in Fischer–Tropsch chemistry, in which the coal is reacted with oxygen and steam at about 1100°C to break up and gasify it into carbon monoxide, hydrogen, and carbon dioxide. A water–gas shift reaction is then carried out to adjust the hydrogen:carbon monoxide ratio of syn-gas, after which the carbon monoxide is catalytically hydrogenated to form methanol or to build up liquid hydrocarbons. Similarly, natural gas can also be used to produce syn-gas. Because the Fischer–Tropsch chemistry is, however, highly energy demanding and uses limited and nonrenewable fossil fuel sources, other approaches are needed for the future.

Since my Nobel Prize in 1994, my interest and significant part of our research effort in the Loker Institute is centered on the chemical reductive recycling of atmospheric carbon dioxide into methyl alcohol and derived hydrocarbons in the contest what we call the "Methanol Economy" (see Chapter 15). This developed over the years to a feasible alternative of oil using essentially CO_2 and H_2 from any available source and energy.

In the quest for an energy sufficient future and fuels, I suggested methanol as the key resource for our fuels, derived products, and materials in an economy named as the "Methanol Economy" (see Chapter 15). Over the last two decades, we have focused on the development of viable and practical method for selective oxidation of methane (natural or shale gas) and effective chemical recycling of carbon dioxide. We developed an efficient *bireforming process* for the exclusive generation of a 1:2 mixture of CO and H_2 which we call as *metgas* from methane for the synthesis of methanol. The "Methanol Economy" and the related work as well as significant enabling discoveries will be mentioned in the ensuing chapters. With Prakash and Goeppert I wrote a monograph *Beyond Oil and Gas: The Methanol Economy* on the sub- • ject which became one of the bestselling books published by Wiley–VCH (third updated edition is under preparation) and was translated into Chinese, Russian, Hungaraian, Japanese, and Swedish languages.

The hydrocarbon research program of the Loker Institute was able in many ways to build on and utilize results of our fundamental work on hydrocarbons and their mechanistic aspects (including carbocationic intermediates) to develop practical processes. The study of such processes became one of the major areas of our research. For example, we developed an environmentally friendly and practical alkylation process for the manufacture of high-octane gasoline using additives that modify the highly volatile and toxic hydrogen fluoride catalyst used industrially and allow its safe use. As mentioned above, we found new ways of hydrocracking coal, shale oil, tar sands, and other heavy petroleum sources and residues using superacidic catalysts. Much work was also done to find new environmentally adaptable and efficient oxygenates for cleaner-burning high-octane gasoline. We also developed improved diesel fuels, making them cleaner burning with high cetane ratings, without the use of toxic additives. Although our ongoing research is also aimed at more efficient utilization of our still-existing fossil fuel resources and development of environmentally benign new chemistry, a major emphasis is to develop new ways to produce hydrocarbons including CO_2 capture and recycling (CCR). This involves new ways of converting methane (natural gas) into higher hydrocarbons and the reductive recycling of excess carbon dioxide (which is at the same time a major greenhouse gas responsible for much of global warming) to useful fuels and products (as discussed in Chapter 13).

Although the major emphasis of the work in the Loker Institute was and is directed in the broadest sense toward the study of the fundamental chemistry of hydrocarbons, substantial and increasing emphasis is also directed to the synthetic aspects of hydrocarbon transformations as well as of derived polymeric and varied synthetic materials.

As a part of a broader hydrocarbon chemistry focus, the study and development of selective *synthetic reagents and methods* are also being pursued. These include organometallic systems, particularly boron- and titanium-based reagents, electrochemical synthesis, and other specific reagent systems for many reactions such as selective alkylation, fluorination, nitration, oxygenation, carbonylation, etc. Emphasis is also given to asymmetric reactions and new catalytic transformations. The synthetic research effort is supplemented by mechanistic studies. In these efforts, my colleague Surya Prakash played an increasingly substantial role and leadership of our endeavors.

My faculty colleagues Hogen-Esch and Weber (now emeritus professor and senior fellow of the Institute) of the Institute brought great expertise in the areas of anionic, cationic, and radical polymerization to the translation of low-molecular-weight hydrocarbons into macromolecules, oligomers, and polymers. Research is also actively pursued in the areas of organosilicon chemistry and the stereochemical and topological control of supramolecular structures.

An integral and equally important part of the Loker Institute's efforts is its educational work. Our research is intimately coupled with the education and development of the next generation of researchers, who will play an essential role in the years to come in helping to solve the very challenging problems our society faces in production and use of hydrocarbons and their products in new, safe, and environmentally adaptable ways. You cannot separate teaching from doing *de facto* research. To facilitate this effort, our faculty, in addition of teaching varied courses and guiding graduate students, also welcomes undergraduates for participation in research projects. Besides our own undergraduates we have, for example, hosted over the years a series of German Adenauer Fellows doing research with us and preparing their undergraduate research theses (diploma work) in the Institute. They all did outstanding job.

We also established an active seminar program and started conducting regular international symposia (once or twice a year) on topics of general interest in our broad field. To date, our Institute has sponsored more than 50 of these highly successful symposia, covering a wide variety of topics. They bring together leading researchers around the world and also allow our faculty and students to have close personal interaction with them. The Stauffer Charitable Trust endowed these symposia some years ago.

· 9 ·

The Nonclassical Ion Controversy and Its Significance

Controversies are an integral part of the progress of science. Many scientific controversies have been well documented. They usually are centered around challenges to established concepts or theories when new facts become available. It took observations using the telescope (in its original primitive but quite effective form) to allow Kepler and Galileo to question the central position of the Earth among the celestial bodies. The theory of the void of space being filled with "ether" succumbed to Michelson and Morley's experiment. The phlogiston theory of chemistry gave away under recognition of the atomic concept of elements. Controversies are eventually resolved because in the physical sciences "proof" (or according to Popper, "falsification") can lead to answers concerning specific questions when new observations or experimental data are obtained, forcing inevitable conclusions. Of course, such fundamental questions as those of our being, the origin of the universe (or universes), of intelligent life, and free will, cannot be similarly approached. Fortunately for chemists, we do not deal with questions reaching outside of the limits of knowledge of our field, and thus our controversies are eventually resolved.

One of the major contemporary chemical controversies in which I was inadvertently increasingly involved developed in the 1950s surprisingly over the structure of a deceiving by simple seven carbon-containing bicyclic carbocation, the 2-norbornyl (bicyclo[2.2.1]heptyl) cation. The

A Life of Magic Chemistry: Autobiographical Reflections Including Post-Nobel Prize Years and the Methanol Economy, Second Updated Edition. George A. Olah with Thomas Mathew.
© 2015 John Wiley & Sons, Inc. Published 2015 by John Wiley & Sons, Inc.

involvement of the ion in the long-recognized rearrangement of nor-bornyl systems prevalent in natural terpenes was first suggested in 1922 by Meerwein. The remarkable facility of skeletal rearrangements in nor-bornyl systems attracted the early interest of chemists. Wagner realized first in 1899 the general nature of these rearrangements and related them to that which takes place during the dehydration of pinacol to tetram-ethylethylene. Sommer found later some tricyclanes in the products of the Wagner rearrangements of terpenes. In 1918 Ruzicka suggested a tricyclane-type mechanism without realizing the ionic nature of the pro-cess. Meerwein reconsidered the mechanism in 1922 and made the far-sighted suggestion that the reaction proceeds through an ionic interme-diate, that is, the norbornyl cation. Hence this type of transformation is now known as the Wagner–Meerwein rearrangement.

The structure of the norbornyl cation became controversial in the "nonclassical ion" controversy following Wilson's original suggestion in 1939 of a mesomeric, σ-delocalized, carbocationic intermediate in the camphene hydrochloride–isobornyl chloride rearrangement. From 1949 to 1952, Winstein and Trifan reported on solvolytic study of the *exo-* and *endo*-2-norbornyl brosylates (*p*-bromobenzenesulfonates) and postulated the intermediacy of a σ-delocalized, symmetrically bridged norbornyl ion intermediate. The *endo* reactant was found to solvolyze in acetic acid, aqueous acetone, and aqueous dioxane to give substitution products of exclusively *exo* configuration, whereas the *exo*-brosylate giving exclusively *exo*-product was markedly more reactive in acetol-ysis than the *endo*, by a factor of 350.

Winstein, one of the most brilliant chemists of his time, con-cluded that "it is attractive to account for these results by way of the bridged (nonclassical) formulation for *the norbornyl cation involving accelerated rate of formation from the exo precursor [by anchimeric assistance]*." His formulation of the norbornyl cation as a σ-bridged species stimulated other workers in the solvolysis field to interpret results in a variety of systems in similar terms of σ-delocalized, bridged carbonium ions.

H. C. Brown (the pioneer of hydroboration chemistry, Nobel Prize, 1979), in contrast, concluded that in solvolysis both 2-*exo* and 2-*endo* norbornyl esters (brosylates, etc.) undergo anchimerically unassisted ionization and that the singular rate and product characteristics of the system are attributable to steric effects, in particular, hindrance to ionization of the *endo* isomers. Explaining the results of the extensive solvolytic studies, he suggested that high *exo/endo* rate and product ratios do not necessitate σ-participation as explanation. In other words *exo* is not fast, *endo* is slow. His suggestion for the structure of the norbornyl cation was that of a rapidly equilibrating pair of regular trivalent ions (classical ions) which he compared to a windshield wiper. However, at the same time none of his studies ever showed that σ-participation is not involved.

In 1962, H. C. Brown initiated his dissent against the σ-bridged 2-norbornyl cation and, for that matter, other nonclassical carbocations. He has maintained his position virtually unchanged over the years and has continued to present his views forcefully. In arguing against carbon σ-bridging, he took the position, despite his pioneering work in structurally closely related boranes, that if carbon were to participate in bridging, novel bonding characteristics must be attributed to it.

In 1965 he stated, "On the other hand, the norbornyl cation does not possess sufficient electrons to provide a pair for all of the bonds required by the proposed bridged structures. *One must propose a new bonding concept, not yet established in carbon structures*" (emphasis added).

In 1967 he again wrote, "The second subclass consists of ions such as the bicyclobutonium and the norbornyl cation in its σ-bridging form, which do not possess sufficient electrons to provide a pair for all of the bonds required by the proposed structures. Therefore, *a new bonding concept not yet established in carbon structures is required*" (emphasis added).

The Brown–Winstein nonclassical ion controversy can be summed up as differing explanations of the same experimental facts (which were repeatedly obtained many times and have not been questioned) of the observed significantly higher rate of the hydrolysis of the 2-*exo* over the 2-*endo*-norbornyl esters. As suggested by Winstein, the reason for this is the σ-participation of the C_1–C_6 single bond leading to delocalization in the bridged "nonclassical" ion. In contrast, Brown maintained that the cause was only steric hindrance to the sterically hindered *endo* side involving rapidly equilibrating "classical" trivalent ions.

Nonclassical ions, a term first used by John Roberts (an outstanding Caltech chemist and pioneer in the field, and a highly respected friend), were defined by Paul Bartlett of Harvard as containing too few electrons to allow a pair for each "bond," that is, they must contain delocalized σ-electrons. This is where the question stood in the early 1960s. The structure of the intermediate 2-norbornyl ion could only be suggested indirectly from rate (kinetic) data and observation of stereochemistry, but no direct observation and structural study was possible at the time.

My own involvement with the norbornyl ion controversy goes back to 1960–1962, when I succeeded in developing a general method of preparing and studying persistent (long-lived) alkyl cations (Chapter 6). Not unexpectedly, my interest extended to the study of various carbocations, including the controversial 2-norbornyl cation. Whereas previous investigators by necessity were able to study carbocations only indirectly (by kinetic and stereochemical studies), my newly discovered methods allowed their preparation and direct study as persistent (long-lived) species.

The memory of the 1962 Brookhaven Mechanism Conference, where I first reported on long-lived carbocations in public, is still clear in my mind. The scheduled "main event" of the meeting was the continuing debate between Saul Winstein and Herbert Brown on the classical or nonclassical nature of carbocations (or carbonium ions as they were still called at the time). It must have come as a surprise to them and to the audience that a young chemist from an industrial laboratory was invited to give a major conference lecture reporting to have obtained and studied stable, long-lived carbonium ions (i.e., carbocations) by the new method of using a highly acidic (superacidic) system. I remember being called aside separately during the conference by Winstein and Brown, both towering and dominating personalities of the time who cautioned me that a young man should be exceedingly careful in making such claims. Each pointed out that most probably I was wrong and could not have obtained long-lived carbocations. Just in case, however, that my method turned out to be real, I was advised to obtain further evidence for the "nonclassical" or "classical" nature (depending on who was giving the advice) of the much-disputed 2-norbornyl cation).

Because my method indeed allowed me to prepare in general carbocations as long-lived species, clearly the opportunity was there to experimentally decide the question through direct observation of the ion. At the time of the Brookhaven conference I had already obtained the proton NMR spectrum of 2-norbornyl fluoride in SbF_5 but only at room temperature, which displayed a single broad peak indicating complete equilibration through hydride shifts and Wagner–Meerwein rearrangement (well known in solvolysis reactions and related transformations of 2-norbornyl systems). However, my curiosity was aroused, and when I moved to Dow's Eastern Research Laboratory in the Boston area in 1964 the work was further pursued in cooperation with Paul Schleyer from Princeton (who became a lifetime close friend) and Martin Saunders from Yale. Using SO_2 as solvent, we were able to lower the temperature of the solution to −78°C, and furthermore, we also prepared the ion by ionization of β-cyclopentenylethyl fluoride or by protonation of nortricylene in $FSO_3H;SbF_5/SO_2ClF$. The three separate routes (representing σ-, π-, or bent σ-participation) gave the identical ion.

I still did not have suitable low-temperature instrumentation of my own to carry out the low-temperature NMR studies, but Martin Saunders at Yale did. Thus our samples now traveled the Massachusetts Turnpike from Boston to New Haven, where with Marty we were able to study solutions of the norbornyl cation at increasingly lower temperatures using his homebuilt variable temperature NMR instrumentation housed in the basement of the old Yale chemistry building. We were able to obtain NMR spectra of the ion at −78°C, where the 3,2-hydride shift was frozen out. However, it took, until 1969, after my move to Cleveland, to develop efficient low-temperature techniques

Figure 9.1. (a) 395 MHz ^{1}H NMR spectra of 2-norbornyl cation in SbF$_5$/SO$_2$ClF/SO$_2$F$_2$ solution. (b) 50 MHz proton decoupled ^{13}C NMR spectra of 2-norbornyl cation (^{13}C enriched) in SbF$_5$/SO$_2$ClF/SO$_2$F$_2$ solution.

using solvents such as SO$_2$ClF and SO$_2$F$_2$. We were eventually able to obtain high-resolution ^{1}H and ^{13}C NMR spectra (using ^{13}C-enriched precursor) of the 2-norbornyl cation down to –159 °C (Figure 9.1). Both the 1,2,6 hydride shifts and the Wagner–Meerwein rearrangement were frozen out at such a low temperature, allowing us to observe the static, bridged ion, which I first reported at the Salt Lake City Organic Symposium in 1969.

The differentiation of bridged nonclassical from rapidly equilibrating classical carbocations based on NMR spectroscopy was difficult because NMR is a relatively slow physical method. We addressed this question in our work using estimated NMR shifts of the two structurally differing ions in comparison with model systems. Later, this task became greatly simplified and more precise by efficient theoretical methods such as IGLO, GIAO, etc., allowing the calculation of NMR shifts of differing ions and comparison with experimental data. It is rewarding to see, however, that our results and conclusions stood up well with all the more recent advanced studies.

We also carried out IR studies (a fast vibrational spectroscopy) early in our work on carbocations. In our studies of the norbornyl cation we obtained Raman spectra as well, although at the time it was not possible to theoretically well calculate the spectra. Comparison with model compounds (2-norbornyl system and nortricyclane, respectively) indicated the symmetrical, bridged nature of the ion. In recent years, Sunko and Schleyer were able, using the since-developed Fourier transform-infrared (FT-IR) method, to obtain the spectrum of the norbornyl cation and to compare it with the theoretically calculated one. Again, it was rewarding that their data were in excellent accord with our earlier work.

Kai Siegbahn's (Nobel Prize in physics, 1981) core electron spectroscopy (ESCA) was another fast physical method that we applied to further resolve the question of bridged versus rapidly equilibrating ions. We were able to study carbocations in the late 1960s by this method, adapting it to superacidic matrixes. George Mateescu and Louise Riemenschneider in my Cleveland laboratory set up ESCA spectroscopy and the necessary methodology for obtaining the ESCA spectra of a number of carbocations, including the *tert*-butyl and the 2-norbornyl cation in SbF_5-based superacidic matrixes. These studies again convincingly showed the nonclassical nature of the 2-norbornyl cation (Figure 9.2a). No trivalent carbenium ion center characteristic of a "classical" ion, such as is the case for the *tert*-butyl cation, was observed in the ESCA spectrum on a timescale where no chemical equilibration process could have any effect. Subsequent ESCA studies by us (with Grunthaner's laboratory at Caltech's Jet Propulsion Laboratory) and by Dave Clark fully justified our previous results and conclusions. So did ever more advanced theoretical calculations (Figure 9.2b).

Additional significant experimental studies were also carried out by others. Arnett reported valuable calorimetric studies. Saunders showed the absence of the deuterium isotopic perturbation of equilibrium expected for a classical equilibrating system. Myhre and Yannoni, at extremely low (5K) temperature, were able to obtain solid-state ^{13}C NMR spectra that still showed no indication of freezing out any equilibrating classical ions. The barrier at this temperature should be as

Figure 9.2. Carbon 1s core-hole-state spectra (1s photoelectron spectra) for the 2-norbornyl cation (a) and Clark's simulated spectra for the classical and nonclassical ions (b).

low as 0.2 kcal/mol (the energy of a vibrational transition). Laube was able to carry out single crystal x-ray structural studies on substituted 2-norbornyl cations. Schleyer's theoretical studies including IGLO and related calculation of NMR shifts and their comparison with experimental data contributed further to the understanding of the σ-bridged carbonium ion nature of the 2-norbornyl cation. The classical 2-norbornyl cation was not even found to be a higher lying intermediate! Very recently the x-ray structure of even the parent 2-norbornyl cation was crystallized at very low temperature and fully substantiated previous conclusions.

As the norbornyl ion controversy evolved, it became a highly public and frequently very personal and bitter public debate. Saul Winstein suddenly died in the fall of 1969, shortly after the Salt Lake City symposium. To my regret, I seemed to have inherited his role to represent the bridged nonclassical ion concept in subsequent discussions. Although

Brown frequently stated that he had not been given the opportunity to express his point of view in public debates, to my recollection, however, a number of these debates took place in which I participated, including meetings at La Grand Motte (France), Bangor (UK), and Seattle, WA.

The 1983 Seattle American Chemical Society Symposium was, in fact, the *de facto* "closing" of the long running debate. Although in the heat of the debates I also made some personal remarks (to my regret, as I expressed subsequently), the experimental evidence at the time was so overwhelming that I concluded my presentation saying, "I don't intend to do any more research on the matter. There is nothing further to be discussed..." My lecture was subsequently published (with Prakash and Saunders) under the title "Conclusion of the Classical–Nonclassical Ion Controversy Based on the Structural Study of the 2-Norbornyl Cation" in an article in the *Accounts of Chemical Research*. In the same issue, Brown wrote, "The nonclassical theory is not necessarily wrong, but it has been too readily accepted" (an obvious understatement concerning the enormous amount of work carried out on the topic). In any case, I kept my promise and have not done further work in the field. The chemical community, and later even the Swedish Academy, accepted the closure of the debate, mentioning inter alia in connection with my Nobel Prize the nonclassical norbornyl ion as one of the resolved carbocation structures.

Before our controversy in 1965 with Herbert Brown and Heinz Staab in Heidelberg

This is the way that the so-called nonclassical ion controversy ended. It basically centered on the question of whether the structure of carbocations, including rapidly equilibrating "classical" ions, can be depicted adequately by using only Lewis type 2-electron 2-center covalent bonding or whether there are also bridged or σ-localized ions, whose structural depiction also necessitates 2-electron 3-center bonding, including higher coordinate carbon. The result was a new understanding of the general bonding nature of carbon compounds as well as of the electron donor ability and reactivity of single bonds in saturated hydrocarbons and σ-bonded compounds in general.

As mentioned, many research groups including ourselves have examined other possibilities and put forward additional proof for the nonclassical bonding phenomena in norbornyl ion. Despite the tedious procedures involved, the low-temperature x-ray crystallographic analysis was suggested to be the best tool to achieve the goal. Very recently, a team of German and American researchers (including my late longtime friend Paul Schleyer) successfully isolated the parent norbornyl cation as the solvated salt $[C_7H_{11}]^+$ $[Al_2Br_7]^-$ •CH_2Br_2 and obtained its structure by x-ray crystallographic analysis. This provided additional unequivocal proof for the bridged nonclassical norbornyl cation.

Intensive, critical studies of a controversial topic also help to eliminate the possibility of any errors. One of my favorite quotations is that by Georg von Békésy, a fellow Hungarian-born biophysicist who studied fundamental questions of the inner ear and hearing (Nobel Prize in medicine, 1961):

> "[One] way of dealing with errors is to have friends who are willing to spend the time necessary to carry out a critical examination of the experimental design beforehand and the results after the experiments have been completed. An even better way is to have an enemy. An enemy is willing to devote a vast amount of time and brain power to ferreting out errors both large and small, and this without any compensation. The trouble is that really capable enemies are scarce; most of them are only ordinary. Another trouble with enemies is that they sometimes develop into friends and lose a good deal of their zeal. It was in this way the writer lost his three best enemies. Everyone, not just scientists, need a few good enemies!"

Clearly there was no lack of devoted adversaries (perhaps a more proper term than enemies) on both sides of the norbornyl ion controversy. It is to their credit that we today probably know more about the structure of carbocations and their bonding nature, such as the norbornyl cation, than about most other chemical species. Their efforts also resulted not only in most rigorous studies but also in the development or improvement of many techniques. Although many believe that too much effort was expanded on the "futile" norbornyl ion controversy, I believe that it eventually resulted in significant insights and consequences to chemistry. It affected in a fundamental way our understanding of the chemical bonding of electron-deficient carbon compounds, extending Kekule's concept of the limiting ability of carbon to associate with no more than four other atoms or groups (see Chapter 10). An equally significant consequence of the norbornyl cation studies was my realization of the ability of saturated C–H and C–C single bonds to act as two electron σ-donors toward strong electrophiles such as carbocations or other highly reactive reagents in superacidic systems, not only in intramolecular but also, as found subsequently, in intermolecular transformations and electrophilic reactions. The key for this reactivity lies in the ability to form two-electron three-center (2e–3c) bonds (familiar in boron and organometallic chemistry). The electrophilic chemistry of saturated hydrocarbons (including that of the parent methane) rapidly evolved based on the recognition of the concept and significance of hypercoordinated carbon, in short, hypercarbon, in chemistry (Chapter 10).

Once the direct observation of stable, long-lived carbocations generally in highly acidic (superacid) systems became possible, it led to my recognition of the general concept of hydrocarbon cations, including the realization that five (and higher coordinate) carbocations are the key to electrophilic reactions at single bonds in saturated hydrocarbons (alkanes, cycloalkanes). In 1972, I offered a general definition of carbocations based on the realization that two distinct classes of carbocations exist (it seemed to be the logical name for all cations of carbon compounds, because the negative ions are called carbanions).

Trivalent ("classical") carbenium ions contain an sp^2-hybridized electron deficient carbon atom, which tends to be planar in the absence of constraining skeletal rigidity or steric interference. The carbenium carbon contains six valence electrons; thus, it is highly electron deficient. The structure of trivalent carbocations can always be adequately described by using only two-electron two-center bonds (Lewis valence bond structures). CH_3^+ is the parent for trivalent ions.

Penta- (or higher) coordinate ("nonclassical") carbonium ions contain five, six, or seven (higher) coordinate carbon atoms. They cannot be described by two-electron two-center single bonds alone but also necessitate the use of two-electron three (or multi)-center bonding. The carbocation center has always eight valence electrons, but overall the carbonium ions are electron deficient because of the sharing of two electrons between three (or more) atoms. CH_5^+ can be considered the parent for carbonium ions.

Lewis' concept that a covalent chemical bond consists of a pair of electrons shared between the two atoms is a cornerstone of structural chemistry. Chemists tend to brand compounds as anomalous whose structures cannot be depicted in terms of such valence bonds alone. Carbocations with too few electrons to allow a pair for each "bond" came to be referred to as "nonclassical," a name first used by Roberts for the cyclopropylcarbinyl cation and adapted by Winstein for the norbornyl cation. The name is still used, even though it is now recognized that, like other compounds, they adopt the structures appropriate for the number of electrons they contain with two-electron two- or two-electron three (even multi)-center bonding not unlike the bonding principle established by Lipscomb (Nobel Prize, 1976) for boron compounds. The prefixes "classical" and "nonclassical," I believe, will gradually fade away as the general principles of bonding are recognized more widely.

Whereas the differentiation of trivalent carbenium and pentacoordinated carbonium ions serves a useful purpose in defining them as limiting cases, it should be clear that in carbocationic systems there always exist varying degrees of delocalization. This can involve participation by neighboring n-donor atoms, π-donor groups, or σ-donor C–H or C–C bonds.

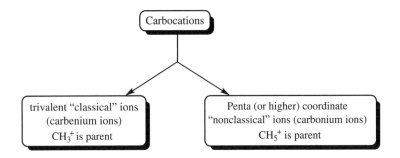

Concerning carbocations, previous usage named the trivalent, planar ions of the CH_3^+ type *carbonium ions*. If the name is considered analogous to other *onium ions* (ammonium, sulfonium, phosphonium ions), then it should relate to the higher valency or coordination state carbocations. These, however, clearly are not the trivalent, but the penta- or higher coordinated cations of the CH_5^+ type. The earlier German and French literature, indeed, frequently used the "carbenium ion" naming for trivalent cations.

Trivalent carbenium ions are the key intermediates in electrophilic reactions of π-donor unsaturated hydrocarbons. At the same time, pentacoordinated carbonium ions are the key to electrophilic reactions of σ-donor saturated hydrocarbons through the ability of C–H or C–C single bonds to participate in carbonium ion formation.

Some characteristic bonding natures in typical nonclassical ions are

Expansion of the carbon octet via 3d orbital participation does not seem possible; there can be only eight valence electrons in the outer shell of carbon, a small first-row element. The valency of carbon cannot exceed four. Kekule's concept of the tetravalence of carbon in bonding terms represents attachment of four atoms (or groups) involving 2e–2c Lewis type bonding. However, nothing prevents carbon from also participating in multicenter bonding involving 2e–3c (or multicenter) bonds (see further discussion in Chapter 10).

Whereas the differentiation of limiting trivalent and penta- or higher coordinate ions serves a useful purpose in establishing the significant differences between these ions, it must be emphasized that these represent only the extremes of a continuum and that there exists a continuum of charge delocalization comprising both intra- and intermolecular interactions.

Neighboring group participation (a term introduced by Winstein) with the vacant p-orbital of a carbenium ion center contributes to its stabilization via delocalization, which can involve atoms with unshared electron pairs (n-donors), π-electron systems (direct conjugate or allylic stabilization), bent σ-bonds (as in cyclopropylcarbinyl cations), and C–H and C–C σ-bonds (hyperconjugation).

Hyperconjugation is the overlap interaction of an appropriately oriented σ-bond with a carbocationic p-orbital to provide electron delocalization with minimal accompanying nuclear reorganization. Nuclear reorganization accompanying σ-bond delocalization can range from little or no rearrangement (hyperconjugation) to partial bridging involving some reorganization of nuclei (σ-participation) and to more extensive or complete bridging. Trivalent carbenium ions, with the exception of the parent CH_3^+, consequently always show varying degrees of delocalization. Eventually in the limiting case carbocations become pentacoordinated carbonium ions. The limiting cases define the extremes of the spectrum of carbocations.

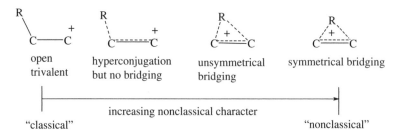

Under superacidic, low nucleophilicity so-called "stable-ion conditions," developing electron-deficient carbocations do not find reactive external nucleophiles to react with; thus they stay persistent in solution stabilized by internal neighboring group interactions.

During my studies I realized that the formation of the σ-delocalized 2-norbornyl cation from 2-norbornyl precursors represented the equivalent of an intramolecular σ-alkylation where a covalent C_1–C_6 bond provided the electrons for the 2e–3c bonded bridged ion (by σ-participation).

In cases of more effective electron donor π- or n-donor neighboring groups, as is the case in forming β-phenylethyl (studied by Don Cram, a good friend from UCLA; Nobel Prize in chemistry, 1987) or β-halogen bridged species, these have sufficient electrons to form 2e–2c bonds (with some intermediate delocalization).

Cram's phenonium ion

ethylene–halonium ion

The intramolecular σ-delocalization in the norbornyl system aroused my interest in studying whether similar electrophilic interactions and reactions of C–H or C–C bonds are possible in intermolecular systems.

This led to my discovery of the general electrophile reactivity of single bonds (Chapter 10). It developed the foundation for the new area of *hypercarbon Chemistry*. The long, drawn-out nonclassical norbornyl ion controversy thus led to an unexpected significant new chapter of chemistry. As frequently happens in science, the drive for understanding (for whatever reason) of what appear at the time to be rather isolated and even relatively unimportant problems can eventually lead to significant new concepts, new chemistry, and even practical applications. It justifies the need for exploration and study in the context of fundamental (basic) research even if initially no practical reasons or uses are indicated. The beauty of science lies in finding the unexpected, and, as Niels Bohr was frequently quoted to have said, "you must be prepared for a surprise," but at the same time you must also understand what your findings mean and what they can be used for. To me, this is the lesson of the norbornyl ion controversy. I strongly believe it was not a waste of effort to pursue it and eventually greatly helped to develop higher coordinate hypercarbon chemistry (Chapter 10).

In 1996, I gave an interview which was recently published in Hargittai's book (*Great Minds, Reflections of 111 Top Scientists* published in 2014). I said at the time answering queries on solving the norbornyl ion controversy...

"I came into it because around 1960 I discovered methods to generate positive organic ions, called now carbocations, as long-lived species, and we were able to take all kinds of spectra and establish their structure, including that of norbornyl cation. In the course of this work I realized, however, that the problem has much wider implications. In the norbornyl ion the C–C single bond acted as an electron donor nucleophile. In this particular case this happens within the molecule, that is intramolecularly. This delocalization which had been originally suggested by Winstein, was indeed there and we were able to see it directly for the first time. Later came what I thougt was a logical idea. The question I asked myself one day was, if this can happen within the molecule, why can't it happen between the molecules? This led to the discovery of a wide range of the electrophilic reactions of saturated hydrocarbons, that is, of C–H or C–C single bonds and the realization that carbon, under some conditions, can indeed bind five or even more neighboring groups."

I still feel the same about the significance of the norbornyl ion controversy today.

· 10 ·

From Kekulé's Four-Valent Carbon to Five- and Higher Coordinate Hypercarbon Chemistry

One of the cornerstones of the chemistry of carbon compounds (organic chemistry) is Kekulé's concept, proposed in 1858, of the tetravalence of carbon. It was independently proposed in the same year by Couper who, however, got little recognition (vide infra). Kekulé realized that carbon can bind at the same time to not more than four other atoms or groups. It can, however, at the same time use one or more of its valences to form bonds to another carbon atoms. In this way carbon can form chains or rings, as well as multiple-bonded compounds.

Kekulé claimed that the concept came to him during a late night ride on a London omnibus in 1854.

> One fine summer evening, I was returning by the last omnibus, "outside"
> as usual, through the deserted streets of the metropolis, which are at other

A Life of Magic Chemistry: Autobiographical Reflections Including Post-Nobel Prize Years and the Methanol Economy, Second Updated Edition. George A. Olah with Thomas Mathew. © 2015 John Wiley & Sons, Inc. Published 2015 by John Wiley & Sons, Inc.

times so full of life. I fell into a reverie and lo! the atoms were gamboling before my eyes…. I saw how, frequently, two smaller atoms united to form a pair, how a larger one embraced two smaller ones; how still larger ones kept hold of three or even four of the smaller; whilst the whole kept whirling in a giddy dance. I saw how the larger ones formed a chain…. I spent part of the night putting on paper at least sketches of these dream forms.

— *August Kekulé, 1890.*

To what degree Kekulé's recollection was factual we do not know, but Couper and Butlerov independently had developed similar, more well-defined concepts of valence bonding, which may have not been entirely unknown to Kekulé.

Another of Kelulé's revelations that supposedly came to him in a dream was his famous structure of benzene. This related to how a carbon chain sometime can close itself into a ring. To satisfy the four valence of carbon, this, of course, raised the need to involve alternating double bonds.

I was sitting writing at my textbook, but the work did not progress; my thoughts were elsewhere. I turned my chair to the fire, and dozed. Again the atoms were gamboling before my eyes. This time the smaller groups kept modestly in the background. My mental eye, rendered more acute by repeated visions of this kind, could now distinguish larger structures of manifold conformations; long rows, sometimes more closely fitted together; all twisting and turning in snake-like motion. But look! What was that? One of the snakes had seized hold of its own tail, and the form whirled mockingly before my eyes. As if by a flash of lightning I woke…. I spent the rest of the night working out the consequences of the hypothesis. Let us learn to dream, gentlemen, and then perhaps we shall learn the truth.

— *August Kekulé, 1865.*

In this case, we now know that the priority for the benzene structure should at least be shared with the Austrian chemist Loschmidt, whose book discussed much of the same concept but preceded Kekulé's. Credit, however, for priority in science, as in other fields, is frequently given to a large degree on the basis of how well the claim became known, how widely it was communicated and disseminated. The longevity and stay-ing power of the claimant also help. Couper's fundamental contribu-tion is not widely recognized, perhaps because he soon thereafter gave up science and never again pursued it. Kekulé's fame and his extensive contributions to chemistry as a leading German professor of his time certainly overshadowed Loschmidt.

Dreams do not come complete with references and credit to preceding work by others. At the same time, it is also true that realizing the signifi-cance of a finding or observation (even if these were originated by some-one else) and applying it to a broad concept of substantial significance and lasting importance is a genuine major contribution to science. Of course, I am not saying that proper credit should not always be given to preceding work or publications, but, regrettably, these sometimes tend to be lost over time. People eventually prefer to quote a single, easily available reference of a paper, book, or review, and even if these contain the original references, in subsequent quotations these are frequently not included. The "dominant" reference thus becomes the sole recognized source.

I certainly do not want to minimize Kekulé's major contributions to chemistry and their significance, but clearly there were—as generally is the case—other contributors who also played a significant role and should be remembered.

Kekulé's four-valent carbon was explained later on the basis of the atomic concept and the "rule of eight" valence electrons of the elec-tronic theory of chemistry. From this, G. N. Lewis introduced the elec-tron pair concept and that of covalent shared electron pair bonding (Lewis bond), which Langmuir (Nobel Prize in chemistry, 1932) fur-ther developed. It was Linus Pauling (Nobel Prize in chemistry, 1954), and others following him, who subsequently applied the principles of the developing quantum theory to the questions of chemical bond-ing. I prefer to use "chemical bonding" instead of "chemical bond," because, after all, in a strict sense the chemists' beloved electron pairs do not exist. Electrons move individually, and it is only the probability that they are found paired in close proximity that justifies the practical term of covalent electron pair bonding. Pauling showed that electron pairs occupying properly oriented orbitals (which themselves are the preferred locations, but do not exist otherwise) cause the tetrahedral structure of methane (involving sp3 hybridization). However, neither Lewis and Langmuir nor Pauling considered that an already shared electron pair could further bind an additional atom, not just two.

In the 1930s, Pauling still believed that diborane has an ethane-like structure and suggested this to Kharash during a visit to Chicago (recalled by H. C. Brown). It was Lipscomb (Nobel Prize, 1976), Pauling's student, who in the 1950s introduced the two-electron three-center (2e–3c) bonding concept into boron chemistry, also explaining the bridged structure of diborane.

It is remarkable that chemists long resisted making the connection between boron and electron-deficient carbon, which, after all, are analogs. The exceptional opportunity was given to me to be able to establish the general concept of five and six coordination of electron-deficient carbon and to open up the field of what I called hypercarbon chemistry.

Organic chemists who are dealing with carbon compounds (or perhaps more correctly with hydrocarbons and their derivatives) have considered 2e–3c bonding limited to some "inorganic" or at best "organometallic" systems and have seen no relevance to their field. The long-drawn-out and sometimes highly personal nonclassical ion controversy was accordingly limited to the structural aspects of some, to most chemists rather obscure, carbocations. Herbert Brown, one of the major participants in the debate and, besides Lipscomb, one of the great boron chemists of our time, was steadfast in his crusade against bridged nonclassical ions. He repeatedly used the argument that if such ions existed, a new, yet unknown bonding concept would need to be discovered to explain them. This, however, is certainly not the case. The close relationship of electron-deficient carbocations with their neutral boron analogs has been frequently pointed out and discussed. Starting in a 1971 paper with DeMember and Commeyras, I pointed out the observed close spectral (IR and Raman) similarities between isoelectronic $^+C(CH_3)_3$ and $B(CH_3)_3$ and emphasized the point repeatedly thereafter. My colleagues Robert Williams, Surya Prakash, and Leslie Field did a fine job in carrying the carbocation, borane, and polyborane analogy much further and also reviewed the topic in depth, for example, in our book, *Hypercarbon Chemistry* (with Kenneth Wade and Árpád Molnár; Second updated edition was also published in 2011).

On the basis of my extensive study of stable, persistent carbocations, reported in more than 300 publications, I was able to develop the general concept of carbocations referred to in Chapter 9. Accordingly, in higher-coordinate (hypercoordinate) carbonium ions of which protonated methane CH_5^+ is parent, besides two-electron two-center Lewis bonding, two-electron three-center bonding is involved.

Extensive *ab initio* calculations, including recent high-level studies, reconfirmed the preferred Cs symmetrical structure for the CH_5^+ cation as we originally suggested with Klopman in 1969. The structure can be viewed as a proton inserted into one of the σ C–H bonds of methane to form a 2e–3c bond between carbon and two hydrogen atoms (or CH_3^+ binding H_2 through a long, weaker bonding interaction). At the same time, we already pointed out that ready bond-to-bond (isotopal) proton migration can take place through low barriers to other possible structures that are energetically only slightly less favorable (which led later to Schleyer's suggestion of a fluxional, completely delocalized structure).

With Lammertsma and Simonetta in 1982, we studied the parent six-coordinate diprotonated methane (CH_6^{2+}), which has two 2e–3c bonding interactions in its minimum-energy structure (C_{2v}). On the basis of *ab initio* calculations, with Rasul we found that the seven-coordinate triprotonated methane (CH_7^{3+}) is also an energy minimum and has three 2e–3c bonding interactions in its minimum-energy structure (C_{3v}). These results indicate the general importance of 2e–3c bonding in protonated alkanes.

$CH_5^+ C_s$ $CH_6^{2+} C_{2v}$ $CH_7^{3+} C_{3v}$

Protonated methanes and their homologues and derivatives are experimentally indicated in superacidic chemistry by hydrogen–deuterium exchange experiments, as well as by core electron (ESCA) spectroscopy of their frozen matrixes. Some of their derivatives could even be isolated as crystalline compounds. In the late 1980s Schmidbaur has prepared gold complex analogs of CH_5^+ and CH_6^{2+} and determined their x-ray structures. The monopositively charged trigonal bipyramidal $\{[(C_6H_5)_3PAu]_5C\}^+$ and the dipositively charged octahedral gold complex $\{[(C_6H_5)_3PAu]_6C\}^{2+}$ contain five- and six-coordinate carbon, respectively. Considering the isolobal relationship (i.e., similarity in bonding) between LAu^+ and H^+, the gold complexes represent the isolobal analogs of CH_5^+ and CH_6^{2+}.

The remarkable stability of the gold complexes is due to significant metal–metal bonding. However, their isolation and structural study are remarkable and greatly contributed to our knowledge of higher coordinate carbocations.

Boron and carbon are consecutive first-row elements. Trivalent carbocations are isoelectronic with the corresponding neutral trivalent boron compounds. Similarly, pentavalent monopositively charged carbonium ions are isoelectronic with the corresponding neutral pentavalent boron compounds. BH_5, which is isoelectronic with CH_5^+, has also C_s symmetrical structure based on high-level *ab initio* calculations. Experimentally, H–D exchange was observed in our work when BH_4^- was treated with deuteriated strong acids, indicating the intermediacy of isopomeric BH_5. The first direct experimental observation (by infrared spectroscopy) of BH_5 has later been reported. The x-ray structure of the five-coordinate gold complex $[(Cy_3P^+)B(AuPPh_3)_4]$ was also reported by Schmidbaur. This square pyramidal compound represents the isolobal analog of BH_5. It further strengthens the relationship of the bonding nature in higher coordinate boron and carbons compounds.

Similarly as five- and six-coordinate CH_5^+ and CH_6^{2+} are isoelectronic with BH_5 and BH_6^+, respectively, seven-coordinate tripositively charged CH_7^{3+} is isoelectronic with the corresponding dipositively charged heptavalent boronium dication $BH_7^{2+.}$ We also have searched for a minimum-energy structure of tetraprotonated methane, CH_8^{4+}. However, CH_8^{4+}, remains even computationally elusive because charge–charge repulsion appears to have reached its prohibitive limit. The isoelectronic boron analog BH_7^{3+}, however, was calculated to be an energy minimum.

As shown in studies of CH_5^+, CH_6^{2+}, and CH_7^{3+} and their analogs, carbon, despite its limiting tetravalence, can still bond simultaneously to five, six, or even seven atoms involving two-electron three (or multi)-center bonds. Such carbon atoms are called *hypercarbons* (short for hypercoordinated carbon atoms).

Because carbon is a first-row element unable to extend its valence shell, *hypervalence* cannot exist in carbon compounds, only *hypercoordination*.

Hypercarbon compounds contain one or more *hypercoordinated* carbon atoms bound not only by 2e–2c but also 2e–3c (or >3c) bonds.

The discovery of a significant number of hypercoordinate carbocations ("nonclassical" ions) initially based on solvolytic studies and subsequently as observable, stable ions in superacidic media as well as on theoretical calculations, shows that carbon hypercoordination is a general phenomenon in electron-deficient hydrocarbon systems. Some characteristic nonclassical carbocations are the following:

Olah
Hogeveen

Sorensen

McMurry

Masamune

Hogeveen

Olah

Pettit
Winstein

Gassman
Winstein
Olah

Winstein
Olah

Schleyer

Masamune
Olah

According to early theoretical calculations, Klopman and I carried out in 1971, the parent molecular ions of alkanes, such as CH_4^+, observed in mass spectrometry also prefer a planar hypercarbon structure.

Even the CH_4^{2+} ion, as calculated later by Radom, has a similar planar C_{2v} structure.

Carbon not only can be involved in a single two-electron three-center bond formation but was shown also capable in some carbodications to simultaneously participate in two 2e–3c bonds. Diprotonated methane

(CH_6^{2+}) and ethane ($C_2H_8^{2+}$), as well as the dimer of the methyl cation ($C_2H_6^{2+}$), are illustrative.

It was the study of hypercarbon containing nonclassical carbocations that allowed us to firmly establish carbon's ability in a hydrocarbon system to bind simultaneously with five (six or even seven) atoms or groups. It should be emphasized that carbocations represent only one class of hypercarbon compounds. A wide variety of neutral hypercarbon compounds, including alkyl (acyl)-bridged organometallics as well as carboranes, carbonyl, and carbide clusters, are known and have been studied. They are reviewed in my book *Hypercarbon Chemistry*. Representative examples are:

[MeBe(C≡CMe)(NMe₃)]₂   2,4-C₂B₅H₇   1,6-C₂B₄H₆

Carboranes

1,12-C₂B₁₀H₁₂   2,3-C₂B₄H₈   Fe₅(CO)₁₅C
[M = Fe(CO)₃]

Carboranes   Metal carbides

Ru₆(CO)₁₇C
M = Ru(CO)₃, M' = Ru(CO)₂)

Rh₆(CO)₁₅C²⁻ (M = Rh(CO)₃)
(all edges CO bridged)

Metal carbides

The most studied hypercoordinate carbocation is the 2-norbornyl cation, around which the nonclassical ion controversy centered (Chapter 9).

The formation of the σ-delocalized norbornyl cation via ionization of 2-norbornyl precursors in low-nucleophilicity, superacidic media, as mentioned, can be considered as an analog of an intramolecular Friedel–Crafts alkylation in a saturated system. Indeed, deprotonation gives nortricyclane.

This realization led me to study related possible intermolecular electrophilic reactions of saturated hydrocarbons. Not only protolytic reactions but also a broad scope of reactions with varied electrophiles (alkylation, formylation, nitration, halogenation, oxygenation, etc.) were found to be feasible, when using superacidic, low-nucleophilicity reaction conditions.

Protonation (and protolysis) of alkanes is readily achieved with superacids. The protonation of methane itself to CH_5^+ as discussed earlier, takes place readily.

Acid-catalyzed isomerization reactions of alkanes as well as alkylation and condensation reactions are initiated by protolytic ionization. Available evidence indicates nonlinear but not necessarily triangular transition states.

The reverse reaction of the protolytic ionization of hydrocarbons to carbocations, that is, the reaction of trivalent carbocations with molecular hydrogen giving their parent hydrocarbons, involves the same five-coordinate carbonium ions.

$$R_3C^+ \ + \ \begin{matrix} H \\ | \\ H \end{matrix} \ \rightleftharpoons \ \left[R_3C\text{-----}\overset{H}{\underset{H}{\diagup}} \right]^+ \ \rightleftharpoons \ R_3C\text{---}H + H^+$$

The isomerization of butane to isobutane in superacids is illustrative of a protolytic isomerization, where no intermediate olefins are present in equilibrium with carbocations.

The superacid-catalyzed cracking of hydrocarbons (a significant practical application) involves not only the formation of trivalent carbocationic sites leading to subsequent β-cleavage but also direct C–C bond protolysis.

$$\begin{array}{ccc} & \left[\overset{R}{\underset{\diagdown}{\diagup C\text{-}C\diagdown}} \right]^+ & \\ \diagup\!\!\!\!\diagdown\!\!+ & \longrightarrow & \longrightarrow \quad R^+ + \quad \diagup\!\!=\!\!\diagdown \end{array}$$

$$\begin{array}{ccc} & \left[\overset{H}{\underset{\diagdown}{-C\quad C-}} \right]^+ & \\ -C\text{-}C- \ + \ H^+ & \longrightarrow & \longrightarrow \quad \overset{+}{\diagup\!\!\diagdown} + \quad \diagup\!\!\diagdown\!\!-H \end{array}$$

Whereas superacid (HF/BF_3, HF/SbF_5, $HF/TaF_5.FSO_3H/SbF_5$, etc.)-catalyzed hydrocarbon transformations were first explored in the liquid phase, subsequently, solid acid catalyst systems, such as those based on Nafion-H, longer chain perfluorinated alkanesulfonic acids, fluorinated graphite intercalates, were also developed and utilized for heterogeneous reactions. The strong acidic nature of zeolite catalysts was also successfully explored in cases such as HZSM-5 at high temperatures.

Not only protolytic reactions but also a whole range of varied electrophilic reactions can be carried out on alkanes under superacidic conditions.

$$R_3C\text{-}H + E^+ \rightleftharpoons \left[\begin{array}{c} R_{\prime\prime\prime} \quad\quad H \\ R\text{-}C\text{----}\langle \\ R^{\nearrow} \quad\quad E \end{array}\right]^+ \begin{array}{c} \longrightarrow R_3C^+ + EH \\ \\ \rightleftharpoons R_3CE + H^+ \end{array}$$

$$R_3C\text{-}CR_3 + E^+ \rightleftharpoons \left[\begin{array}{c} R_{\prime\prime\prime} \quad\quad CR_3 \\ R\text{-}C\text{----}\langle \\ R^{\nearrow} \quad\quad E \end{array}\right]^+ \rightleftharpoons R_3CE + R_3C^+$$

$$E = D^+, H^+, R^+, NO_2^+, Hal^+, HCO^+, \text{etc.}$$

Alkylation of isoalkanes with alkenes is of particular significance. The industrially used alkylation of isobutane with isobutylene to isooctane is, however, *de facto* alkylation of the reactive isobutylene and not of the saturated hydrocarbon. Isobutane only acts as a hydride transfer agent and a source of the *tert*-butyl cation, formed via intermolecular hydride transfer. In contrast, when the *tert*-butyl cation is reacted with isobutane under superacidic conditions and thus in the absence of isobutylene, the major fast reaction is still hydride transfer, but a detectable amount of 2,2,3,3-tetramethylbutane, the σ-alkylation product, is also obtained. With sterically less crowded systems σ-alkylation becomes more predominant.

A fundamental difference exists between conventional acid-catalyzed and superacidic hydrocarbon chemistry. In the former, trivalent carbenium ions are always in equilibrium with olefins, which play the key role, whereas in the latter, hydrocarbon transformation can take place without the involvement of olefins through the intermediacy of five-coordinate carbocations.

The reaction of trivalent carbocations with carbon monoxide giving acyl cations is the key step in the well-known and industrially used Koch–Haaf reaction of preparing branched carboxylic acids from alkenes or alcohols. For example, in this way, isobutylene or *tert*-butyl alcohol is converted into pivalic acid. In contrast, based on the superacidic activation of electrophiles leading to superelectrophiles (see Chapter 12), we found it possible to formylate isoalkanes to aldehydes, which subsequently rearrange to their corresponding branched ketones. These are effective high-octane gasoline additive oxygenates. The conversion of isobutane into isopropyl methyl ketone, or isopentane into isobutyl methyl ketone is illustrative. In this reaction, no branched carboxylic acids (Koch-products) are formed.

$$(CH_3)_3CH \xrightarrow[\text{HF/BF}_3]{\text{CO}} \left[(CH_3)_3C \overset{H}{\underset{\diagdown}{\diagup}} CHO \right]^+$$

$$(CH_3)_2CHCOCH_3 \xleftarrow{\text{HF–BF}_3} (CH_3)_3CCHO$$

The superacid-catalyzed electrophilic oxygenation of saturated hydrocarbons, including methane with hydrogen peroxide (via $H_3O_2^+$) or ozone (via HO_3^+), allowed the efficient preparation of oxygenated derivatives.

$$H_2O_2 \xrightarrow{H^+} \overset{H}{\underset{H}{\diagdown}} \overset{+}{O}-OH \xrightarrow{CH_4} \left[H_3C \overset{H}{\underset{OH}{-\diagdown}} \right]^+ \longrightarrow CH_3OH_2^+ \xrightarrow{-H^+} CH_3OH$$

$$O_3 \xrightarrow{H^+} O\text{-}O\text{-}\overset{+}{O}H \xrightarrow[\text{CH}_4]{-O_2} \left[H_3C \overset{H}{\underset{OOOH}{-\diagdown}} \right]^+ \xrightarrow{-H_2O_2}$$

$$CH_2O \xleftarrow{-H^+} \overset{+}{CH_2OH} \longleftarrow \left[CH_3O^+ \right]$$

Because the protonation of ozone removes its dipolar nature, the electrophilic chemistry of HO_3^+, a very efficient oxygenating electrophile, has no relevance to conventional ozone chemistry. The superacid-catalyzed reaction of isobutane with ozone giving acetone and methyl alcohol, the aliphatic equivalent of the industrially significant Hock-reaction of cumene is illustrative.

Electrophilic insertion reactions into C–H (and C–C) bonds under low-nucleophilicity superacidic conditions are not unique to alkane activation processes. The C–H (and C–C) bond activation by organometallic complexes, such as Bergman's iridium complexes and other transition metal systems (rhodium, osmium, rhenium, etc.), is based on somewhat similar electrophilic insertions. These reactions, however, cannot as yet be made catalytic, although future work may change this. A wide variety of further reactions of hydrocarbons with coordinatively unsaturated metal compounds and reagents involving hypercarbon intermediates (transition states) are also recognized, ranging from hydrometallations to Ziegler–Natta polymerization.

In the conclusion of my 1972 paper on the general concept of carbocations, I wrote, "The realization of the electron donor ability of shared (bonding) electron pairs, including those of single bonds, should rank one day equal in importance with that of unshared (nonbonding) electron pairs recognized by Lewis. We can now not only explain the reactivity of saturated hydrocarbons and in general of single bonds in electrophilic reactions, but indeed use this understanding to explore new areas of carbocation chemistry."

This was one of the few times I ever made a prediction of the possible future significance of my chemistry. Close to half a century later, I take some satisfaction that I was correct and that, indeed, hypercarbon chemistry has a significant place on the wide palette of chemistry.

Carbon can extend its bonding from Kekulé's tetravalent limit to five-, and even higher bonded (coordinate) hypercarbon systems. Higher coordinate carbocations are now well recognized. They are the key to our understanding of the electrophilic reactivity of C–H or C–C single bonds and of hypercarbon chemistry in general. Some of their derivatives, such as some gold complexes, can even be isolated as stable crystalline compounds. The chemistry of higher coordinate carbon (i.e., hypercarbon chemistry) is rapidly expanding, with many new vistas to be explored. The road from Kelulé's tetravalent carbon to hypercarbon chemistry took a century to travel. Carbon also unveiled other unexpected new aspects, for example, fullerene-type allotropes and their chemistry. There is no reason to believe that our century will not bring much further progress in the hypercarbon chemistry. The further functionalization and conversion of methane (natural gas) is already widely studied and developed. Hypercarbon chemistry also played a significant role in the development and enabling the concept of the "Methanol Economy" (Chapter 15).

· 11 ·

The Nobel Prize:
Learning to Live with It and Not Rest on Laurels

Early in the morning on October 12, 1994, the phone rang in our home in Los Angeles on what promised to be another wonderful, sunny California day. Both my wife and I are early risers, and at 6 A.M. as usual I was already well into my morning routine having breakfast and preparing for my day at the university. We had been away for a few days to the San Francisco area and had returned the previous afternoon, so I had a busy schedule ahead of me. The voice on the phone was that of the secretary of the Royal Swedish Academy, and the call changed my plans for the day and, in many ways, my relatively quiet, well-organized life. He informed me that the Academy had just voted to award me the 1994 Nobel Prize in chemistry and asked whether I would accept it (a question I believe recipients do not find difficult to answer). He then added his own congratulations and in good humor suggested that to establish the bona fide nature of his call and that he was not a prankster, he put on some of my Swedish friends who were standing by and whose voices I probably would recognize. I indeed did and was grateful for their good wishes.

My day, which had started out in its regular routine, from this point on became quite hectic. I barely had time to call my sons, who for some years at this time in October used to make "polite" and "understanding" remarks to the fact that their father was again not on the list of Nobelists (not that anyone really can or should expect this). I also called a few

A Life of Magic Chemistry: Autobiographical Reflections Including Post-Nobel Prize Years and the Methanol Economy, Second Updated Edition. George A. Olah with Thomas Mathew.
© 2015 John Wiley & Sons, Inc. Published 2015 by John Wiley & Sons, Inc.

close friends, including my colleague Surya Prakash, as well as Katherine Loker and Carl Franklin, who had greatly helped my work since I moved to California and had always great faith in me. From that point on our phone never stopped ringing. Amazingly, within a short while, even before I was able to shave and get fully dressed, some reporters started to show up at our house, which is on a remote cul-de-sac of a canyon on the top of Beverly Hills. By the time I finished my by then cold morning coffee I was giving the first of many interviews to a reporter from the Voice of America, who was fastest in finding our home. One of the first questions he asked (to be repeated many times) was what I was going to do with the prize money. My spontaneous answer was that I was going to give the check to my wife, who handles our financial affairs.

The rest of the day was equally hectic. By the time Judy and I made it to the university, my colleagues and students had already organized a most touching welcome. I overheard my graduate student Eric Marinez proudly declaring "we won," reflecting the close-knit spirit of my group and perhaps the USC spirit of athletic competitiveness (which I must confess, however, had never enticed me). This was followed by a reception given by the President of the University, Steven Sample, who always was very supportive of my work and who made gracious remarks. Other events with much press and TV coverage followed. There was, understandably, substantial excitement on the USC campus because at the time I was the first Nobel winner in the University's 114-year history. Attention on the campus was shifted at least for the time (but I hope also for the future) to academic achievements matching the long-standing recognition of USC as a premier athletic institution with more Olympic medal winners than any other US university and an outstanding record in football, baseball, track, and other sports. I am glad to mention that 19 years after my Nobel win, my friend and colleague Arieh joined me in 2013 as USC's second Laureate in Chemistry.

Was I prepared for the Nobel Prize? I probably was, because over the years friends and colleagues had hinted that I had been nominated many times (of course, nobody really "knows" anything about the selection in advance). It is frequently said that the Nobel Prize represents the *de facto* end of the active research career of the recipients and they become public figures with little time left for scholarly work. However, I was determined that no prize or recognition would substantially change my life (see also Chapter 16). Furthermore, Judy, my life partner and strength in all situations, never would have stood for changes affecting the real values we had come to believe in and the life we had built for ourselves.

The Olah group, the day my Nobel Prize was announced with friends Bob Williams, Surya Prakash, Joe Casanova, and Reiko Choy, my secretary

Living with the Nobel Prize in many ways is not easy, not the least because it involves new obligations, public appearances, speeches, etc. However, it never brought any real problems to our life and relationship with our family, including my scientific family of more than 200 former students and postdocs, our friends, and colleagues. We did learn, nevertheless, to differentiate friends and acquaintances as well as public recognition (such as honorary degrees, memberships in academies, societies, etc.) dating before and after the Nobel Prize. The latter, I feel, often mean more interest in Alfred Nobel's prize than in the individual who won it. It is useful to keep this in mind before being carried away by one's instantly acquired importance as well as the assumed universal wisdom and knowledge of all topics that Nobelists are frequently asked about.

With Judy and sons George and Ron, and daughter-in-law Cindy at the USC reception after the announcement of the prize, October 12, 1994

It is clear that a physicist, chemist, etc., cannot be an "expert" even in his broad and complex field of science, not to mention topics outside of science. Becoming a universal oracle of wisdom on all matters is something I was never tempted in the slightest to accept. It is perhaps a reflection on our times and society that publicity seems to be important for many regardless of the context in which they are mentioned. I certainly do not want to criticize others whose views are different from mine, but maintaining self-respect and a healthy appreciation of one's limitations and relative significance is very useful. I believe that I am reasonably well educated and well read in many areas outside of my field, and I have certainly always had strong views. Not surprisingly, few, if any, gave attention to these views before my Nobel (except my long-suffering wife who, over the years, must have heard some of them more often than she would like to remember). It was thus amusing to observe that after you receive the prize people start to believe that you are worth listening to.

I have many other recollections of the hectic days following the announcement of my prize. The flow of good wishes and personal messages was overwhelming. I made a firm commitment to answer them all in a personal way. We worked out an efficient system to do this, and with the help of my wonderful secretary, Reiko Choy, we accomplished it. I remember Bob Woodward many years ago took his secretary to Stockholm to answer good wishes directly from there, making it even more personal. I had not gone to this extreme, but answering properly and promptly was important to me, even more so because I had never gotten used that some people never answer you or return your calls, not necessarily to disregard you, but perhaps because they are too busy or feel it is unimportant. In this regard, too, I can say that winning a Nobel Prize brings about changes. Your letters are always answered, and your phone calls are promptly returned.

Winning the Nobel Prize brings with it many invitations for various events. One of the first invitations we received with my wife was from the President and Mrs. Clinton to come to the White House for a celebration of the year's American Nobel Prize winners. We had been to the White House before only as tourists guided through some of the public areas. As first-generation immigrants we certainly viewed the invitation as an honor, and we went to Washington with great anticipation. The event was indeed most impressive. It was also a pleasure to meet my fellow laureates, with whom we spent subsequently a memorable week in Stockholm. The President himself never showed up (we were told that he was otherwise very busy that day), but Mrs. Clinton and Vice President Gore were hosting the event. I was particularly impressed by Mrs. Clinton (although I must admit I do not share her politics). She spoke for 15 minutes quite impressively about the significance of science to the audience of some 150–200, including cabinet secretaries, members of Congress, directors of agencies, representatives of various scientific organizations, and invited various guests, without any notes or a teleprompter. The Vice President, in contrast, followed by reading a prepared text. He reflected on the impressive record of American Nobel winners. He mentioned that on the mantelpiece in one of the rooms of the White House there is the Nobel medal of President Theodore Roosevelt, the first American Nobel Peace Prize winner (for settling the Spanish War). On a lighter note, he apologized for the rather extensive security procedures involved for guests entering the White House. He also recalled a story of the Swedish king who, while visiting Germany, was congratulated on having such a famous chemist as Scheele (one of the "inventors" of oxygen) in his country. Neither the king nor anybody in his entourage had any idea who Scheele was. Nevertheless, the king gave an order to find him and, if he was a "decent fellow," to make him a count. Months later, it was reported to him that Scheele was found and, being a cavalry officer, was considered a reliable man and consequently honored. As it turned out, however, he was the wrong Scheele, having nothing to do with science or chemistry. Gore concluded his story that strict White House security also helps to prevent to admit the wrong scientists to events such as this.

Much has already been written about the Stockholm Nobel ceremonies. I would like, therefore, only to mention some personal recollections. The events, like all the activities of the Nobel Foundation, are superbly organized. You get specific information covering all details in advance. The protocol assigns to each prize the privilege of inviting 10 guests to the major ceremonies and events. Because I won the chemistry prize alone (prizes can be divided between no more than three persons), I could invite all 10 guests myself (otherwise, like the prize money, the invitations are also divided). We took our sons George and Ron and our daughters-in-law Sally and Cindy with us and invited Judy's cousin Steve

White House reception, November 1994, with Vice President Al Gore

Peto and his wife Magda, who lived in Brussels. Two close friends and colleagues, Surya Prakash and Peter Stang, also joined us. Katherine Loker (a great benefactor whose name our Institute bears) and Carl Franklin also accompanied us. Carl's late wife Caroline used to tease me that if I ever got to Stockholm they would come along. I am sure in spirit she was there beside Carl.

The Nobel Prizes are awarded each year on December 10, the anniversary of Alfred Nobel's death. The preceding week is taken up by formal events and lectures by the laureates. The fairy tale week had already started on our SAS flight to Stockholm, when the captain greeted us on the public address system. At the airport there was a reception, and then we were driven in a limousine, which was at our disposal during our whole visit, to the famous Grand Hotel. We were accompanied by our "attaché" Ms. Carina Martensson from the Swedish Foreign Office, who became our "guardian angel."

With our guardian angels in Stockholm (attaché Carina Martensson with the briefcase)

She kept everything running smoothly and always got us to our many appointments on time. She had attended the Monterey Institute of International Studies in California and was well acquainted with many California institutes, including Berkeley, Stanford, and Caltech (USC, however, must have been a novelty to her, too.) It is a tradition that the Swedish Foreign Ministry assigns each laureate an attaché, which we learned is a prestigious and much sought after assignment, generally signaling a promising career. The Grand Hotel where we were put up traditionally houses the Nobelists. In earlier times when the events were much smaller, even the ceremonies were held there. These have since moved to the Concert Hall and Stockholm City Hall.

Alfred Nobel died in San Remo, Italy in 1896, and his will established the Prizes in physics, chemistry, physiology or medicine, literature, and the Peace Prize given in Oslo (Norway). They were first given in 1901, and they now span a century. In 1968, the Bank of Sweden established an additional Alfred Nobel Memorial Prize in Economic Sciences, which although not one of the original Nobel Prizes, is given out together with them and for all practical purposes has become the Nobel Prize in economics.

Whereas the events of the Nobel week are well reported, it is difficult to describe the spirit of all of Sweden during these events and their general meaning to the Scandinavian people. This struck me particularly during an event we attended honoring three high school science teachers selected for a special award for outstanding teaching. The event was televised all over Scandinavia and must have been most inspirational for students and their parents alike. In general, the importance of education and the significance of science for the future of mankind were emphasized throughout the week in a way which is unimaginable for Americans, who never see anything like this on their TV screens (and would never stand for it to replace their favorite shows and football games). Alfred Nobel's heritage thus has taken up national significance and acts as an inspiration for the whole country.

During our stay in Stockholm we attended the traditional Nobel concert in the famous Concert Hall (where later the awards themselves were presented). It was impressive to see the informality of the event. No particular security was evident for the Royal couple, who freely mingled with the invited guests and audience.

A luncheon for the American laureates and some 200 guests was given at the residence of Thomas Siebert, the American Ambassador to Sweden. The Ambassador, a former university classmate of President Clinton, and his wife were most gracious. We were also invited by the Hungarian Ambassador to a reception given for the two Hungarian-born winners (the economist John Harsanyi and myself). Little Hungary indeed can be proud to have produced such an impressive number of Nobel winners in proportion to its population (although nearly all emigrated and did most of their work abroad). Some believe that there must be some special talent in the Hungarians for certain fields (besides the sciences, music, film making, engineering, and entrepreneurship are frequently mentioned.) I feel, however, that the main reason was a good educational system. This is a more realistic explanation than is sometimes offered to explain the success of Hungarian-born scientists. During the Manhattan project, in which the Hungarian-born Neuman, Szilard, Teller, and Wigner played an important role, Enrico Fermi was quoted as suggesting that they were really visitors from Mars. They possessed advanced intelligence, but found themselves in difficulty because they spoke English with a bad foreign accent, which would give them away. Therefore, they chose to pretend to be Hungarians, whose inability to speak any language but Hungarian without a thick foreign accent is well known (I am myself a good example of this). The story was memorialized in the book *Voice of the Martians* by the Hungarian physicist George Marx. It is an attractive saga, but in fact, the development of scientists depends to a great degree on a good education, which should be always emphasized.

Another event that stands out in my memory was the Nobel Lecture I gave before the Swedish Academy of Sciences, chaired by Professor Kerstin Fredga, its President and the leading Swedish space scientist. She is the daughter of the late Arne Fredga, a chemistry professor and longtime member of the Nobel Committee and the Nobel Foundation. I had known him and visited him in Uppsala years before, thus it was even more of a personal pleasure to meet his daughter. The only formal obligation connected with the Nobel Prize is to deliver this lecture in Stockholm. It is a distinct honor, and these lectures together with some biographical material are published in the annually published volumes *Les Prix Nobel*.

One evening during that week our friends Lars and Edith Ernster invited our family and some mutual friends for dinner. Lars (or Laci, as a born Hungarian) was a distinguished professor of biochemistry at Stockholm University who, regrettably, passed away recently. His wife Edith, a talented violinist, was the first female concert master of the Stockholm Opera Orchestra. I do not know how they managed the dinner for all of us in their apartment, but it was a most pleasurable and relaxing evening. The night before the award presentations, we ourselves gave a dinner in a restaurant for our friends and guests, which also came off well.

Saturday, December 10, finally arrived. In the morning, there was a rehearsal of the award ceremony (with a stand-in for the King). Instead of the formal tails obligatory at the ceremony itself, I could wear a warm

Nobel Lecture, Stockholm, December 1994, with Professor Fredga, President, Royal Swedish Academy of Science

sweater and slacks, much more familiar and comfortable attire. We were instructed in detail and rehearsed the event. The award ceremony started on a dark, cold afternoon at exactly at 4:30 P.M. (punctuality, we learned, is obligatory in Sweden even from the King himself). With musical accompaniment and regal pomp, the ceremony proceeded with great precision. There is a time-honored sequence in which the prizes are presented, starting with physics, followed by chemistry, physiology or medicine, literature, and then economics (the Peace Prize is presented in Oslo on the same day).

When my turn came, I listened as Professor Salo Gronovitz, the Chairman of the Chemistry Nobel Committee, introduced me to the King and Queen and the formal assembly, including some former prize winners, members of the Swedish Academies, and visiting guests. He spoke initially in Swedish but ended up with the following in English: "Professor Olah, I have in these few minutes tried to explain your immense impact on physical organic chemistry through your fundamental investigations of the structure, stability, and reactions of carbocations. In recognition of your important contribution, the Royal Swedish Academy of Sciences has decided to confer upon you this year's Nobel Prize for chemistry. It is an honor and pleasure for me to extend to you the congratulations of the Royal Swedish Academy of Sciences and to ask you to receive your Prize from the hands of His Majesty, the King." With this, I stepped into the circle on the carpet marked with a large N and received my prize. The significance of the moment all at once struck me. The long journey that started in Budapest flashed through my mind and also some doubt about whether all these were real and were *de facto* happening. I forced myself, however, to the present and accepted my medal and the impressive diploma, which on one side has a watercolor painting, as I learned later, by the artist Bengt Landin.

After completion of the award ceremony in the Concert Hall we were driven to the city hall for the Nobel banquet. By this time it was a bitterly cold, windy evening. The courtyard of the city hall was lighted up by the torches of hundreds of school children lining it. It was a most impressive sight, but we felt sorry for the children, who must have braved the weather for a long while.

Receiving the Nobel Prize from King Carl Gustav XVI, December 10, 1994

The Nobel banquet in the Blue Hall of the Stockholm City Hall is traditionally attended by in addition to the Swedish Royal Family, members
of the Government and Parliament, diplomats, and some 1200 invited
guests. Attendance of the banquet, we were told, is the most sought-after
social invitation in Sweden.

We were first taken to a suite for brief cocktails and to get organized for the entry into the banquet hall. The procession formed and
then began. It was led by King Carl Gustaf XVI with Judy on his arm.
This all followed protocol, according to which the King accompanies
the wife of the physics winner, if there is a sole winner, or if this is not
the case that of the chemistry winner, etc. Because the physics prize was
shared that year, but I was the sole chemistry winner, Judy found her-
self leading the procession on the arm of the King. They were followed
by the Queen, escorted by the President of the Nobel Foundation. The
procession started on the mezzanine floor, which ran the length of the

hall, and we marched down the impressive marble stairs into the large ground floor hall and then split to either side of a very long table. It was a most thrilling entrance, with trumpets blaring and everyone standing and cheering.

At the dinner, too, Judy was seated next to the King, and I was across the table at the side of the Queen. Queen Silvia is a beautiful and gracious lady. During the dinner I learned she was born in Heidelberg, Germany; her mother was from Brazil, where she spent part of her youth. We talked about many topics, not the least our children and her many duties, including the annual Nobel events, with which by now she was very familiar. The guests were seated at tables perpendicular to the main one, with our family and guests sharing a table with our Swedish chemist colleagues.

The banquet started with the Chairman of the Foundation proposing a toast to the King, who then answered with a silent toast to the memory of the great benefactor and philanthropist Alfred Nobel.

The banquet itself was a superbly choreographed event. Each course was brought in by hundreds of waiters descending the same staircase with musical and lighting accompaniment like a well-orchestrated ballet. The tables were beautifully set with gold-plated china and crystal specially designed and used only for the Nobel banquet and decorated with a profusion of flowers. These, I learned, were flown in from San Remo in Italy, where Alfred Nobel had a winter home (and where he died).

Toward the end of the dinner laureates representing each prize were called upon for a brief (not more than 3 minutes) remark. I generally do not write down any of my remarks or speeches, but this was a special occasion and I was requested to provide a text. Thus I can reprint here what I said:

> Your Majesties, Your Royal Highnesses, Ladies. and Gentlemen, I am most grateful for the honor bestowed on me today. Although receiving the Nobel Prize is the greatest satisfaction any scientist can experience, I consider it not only a personal acknowledgment, but also that of all my students, associates, and colleagues whose dedicated work over the years allowed my field of chemistry, which is not frequently highlighted in public, to be recognized.

There are many facets of chemistry. Mankind's drive to uncover the secrets of live processes and use this knowledge led to spectacular advances in the biological and health sciences. Chemistry richly contributes to this by helping our understanding at the molecular level. Chemistry is, however, and always will be a central science of its own.

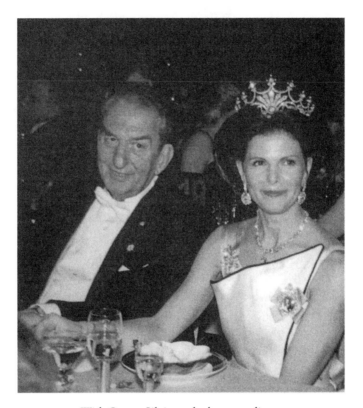

With Queen Silvia at the banquet dinner

Chemists make compounds and strive to understand their reactions. My own interest lies in the chemistry of the compounds of the elements carbon and hydrogen, called hydrocarbons. These make up petroleum oil and natural gas and thus are in many ways essential for everyday life. They generate energy and heat our houses, fuel our cars and airplanes and are raw materials for most man-made materials ranging from plastics to pharmaceuticals. Many of the chemical reactions essential to hydrocarbons are catalyzed by acids and proceed through positive ion intermediates, called carbocations.

Remarks at the Nobel dinner

To be able to prepare and study these elusive species in stable form, acids billions of times stronger than concentrated sulfuric acid were needed (so-called superacids). Some substituted carbocations, however, are remarkably stable and are even present in nature. You may be surprised to learn that the fine red wine we drank tonight contained carbocations which are responsible for the red color of this natural 12% or so alcoholic solution. I hope you enjoyed it as much as I did.

Chemistry does not always enjoy the best of reputations. Many of our plants and refineries are still potentially dangerous and may pollute their surroundings. At the same time, our society enjoys a high standard of living not in small measure through the results of chemistry, which few would give up. I believe that chemistry can and will be able to bring about an equilibrium between mankind's needs and our environmental concerns. Chemistry will continue to benefit mankind in the spirit of Alfred Nobel, a fellow chemist whose example continues to inspire us all.

The audience particularly liked the red wine part. So did I, because usually it is not easy to explain what my "carbocations" are all about.

At the end of the banquet, we ascended the stairs to the Gold Room, where there was dancing till the early hours of the next day. Judy and I were, however, too exhausted after the long day, and after some private time with our Royal hosts we returned (or shall I say floated back) to our hotel.

The next day, Sunday the 11th, the Royal couple gave a dinner in their palace for the laureates and some 200 guests as a conclusion of the Nobel week. The austere, impressive palace I was told by the Queen, is not suited to bringing up a family and they use it only for formal entertainment.

There was one other event for us to remember. As December 10 is Nobel day in Sweden, December 13 is St. Lucia's day. We were awakened early morning by a knock on our door and we were greeted by a singing group of white-clad girls carrying candles and a traditional breakfast to honor Santa Lucia. It was a moving and memorable morning.

I should mention that in 2001 we attended the centennial celebration of the Nobel Prizes in Stockholm. The spectacular event brought back many fine memories.

With the King and Queen of Sweden after the Nobel banquet

During the week in Stockholm I also visited the University of Stockholm and the Royal Technological Institute. Subsequently, I also visited and lectured at Uppsala University and the Universities of Gothenburg and Lund. We were received everywhere with great friendship and hospitality ending our most memorable trip in Copenhagen, from where we flew home. It was time to come back down to earth from the skies and resume our life.

The Nobel Prizes also come with a monetary award, which that year amounted to close to a million dollars. After paying taxes on it (the United States is the only country that taxes the Nobel Prizes), we donated part of it to help endow a chair in chemistry at USC (which was later awarded to my friend and colleague Surya Prakash). We also donated endowment for a prize for younger chemists in my native Hungary. The balance was shared with our children. It was thus not difficult to dispose of the prize money, but money, of course, is really not the essential part of the Nobel.

Concerning taxes, I was told that even the Nobel gold medal should be taxed based on its weight (seemingly the only value tax authorities put on it). I must confess, however, not to have paid duty on it, even after declaring it properly upon our return home. The nice lady custom inspector inquired us how I acquired a gold medal. My wife told her that it was the Nobel medal I had just received in Stockholm. To my surprise, she not only knew what this was but shook my hand, saying that I was the second Nobel laureate she had had the pleasure to meet personally (the other was Linus Pauling). She decided that my "acquisition" of the medal did not come under the rules to pay duty. Fortunately, neither of us get into any trouble for it.

Winning the Nobel Prize inevitably brings with it, besides a brief period of wider publicity (which in America mercifully evaporates fast), a steady stream of invitations, varied honors and recognitions, as well as more general public involvement. Professors and scientists in American life are usually not exactly at the top of the "social ladder," nor are they used to much recognition. Personally, I rather like this, because it helps not to attach overgrown significance to one's importance, keeps one humanized, and, most importantly, allows one to stay centered without much distractions on work. It was, therefore, certainly only a passing shock that for a while the limelight of the Prize seemed to intrude on my well-organized, quiet life. However, friends who have gone through this before assured me that this would soon slow down and that if you are set to continue your life in your "own way," it is possible to manage it. They were indeed right.

Being fairly strong willed, I was determined to continue to live my life with as few distractions as possible. Thus I was back at my university duties the day after our return from Stockholm. Research and teaching for me was always an integral part of my life, not just a "work habit," and I could not see any reason for change. I also did not feel ready for retirement, formal or *de facto* while pretending otherwise. I like teaching and still feel able to contribute significantly to my research. In short, I felt disinclined to rest on past laurels.

It was easier to set my goals than to carry them out. All of us have only a finite amount of energy and available time. Furthermore, one of the most important aspects of my life always was to keep a balance between my professional and personal life. Even so, I am afraid I had frequently shortchanged my family particularly my children. Under no circumstances was I prepared to do this again, with our adorable grandchildren Peter, Kaitlyn, and Justin also added new vistas and pleasures to our family life. On the other hand, I believe it is essential to stay active, but outside chemistry I know little ways to achieve this myself. However, I did make adjustments. I decided not to take on new graduate students, as the long-range major commitment needed to properly guide them through 4 or 5 years of research seemed unrealistic. I continued, however, to work closely with graduate students already in my group, helping and counseling students of the Loker Institute and work with a limited number of my postdoctoral fellows, who became the mainstay of my research group. My close cooperation with my colleague and friend Surya Prakash continued unabated in areas of mutual interest. He built up over the years an impressive independent research program on his own, while in areas of hydrocarbon chemistry and some other fields of mutual interest our cooperative joint research continued. It also extended into the exciting new areas of the "Methanol Economy" which will be discussed in the ensuing Chapters 13–15.

One of the essentials of living with the Nobel Prize was to learn to firmly say *no* to many new commitments. Of the large number of invitations I receive for various events, recognitions, lectures, etc., I only accepted a few. In any case, many of these are frequently not so much for me personally as for the lure of having a Nobel Prize winner for the event. Once you realize this, it is easier to decline. For a long time, I practically never traveled without Judy and now do not travel anywhere. With passing years we gave up travel altogether. For us there is no place to live like Southern California. We still swim till very recently in our pool, to keep in shape and for our health. Where else could you do this, and afterwards enjoy your chemistry?

As long as the enjoyment of leading our research seminars, doing research, and working with my younger associates and colleagues lasts and I feel able to make meaningful contributions, I intend to continue and not to rest on any past achievements. I believe this also keeps me active and interested past an age when most others would have decided long ago to quit. Otherwise, continuing what I enjoy doing still comes naturally, and I still have creative ideas and follow them up. One day when this will not be the case anymore, I fully intend to retire and start to act my age. I received in 2001, the Arthur C. Cope Award by the American Chemical Society which emphasizes achievements, the significance of which became apparent in the years after my Nobel Prize and later in 2005 the Priestley Medal, the highest honor of the American Chemical Society. In 2013, I was awarded together with Surya Prakash the inaugural Eric & Sheila Samson Prize of the Prime Minister of Israel for our work for alternative fuels and the Methanol Economy (Chapter 15). These give me particular pleasure because they acknowledged my work done after my Nobel Prize. Otherwise, I try to follow the advice of a late friend, Jay Kochi, who sent me a quotation from Edward Lavin's book *Life Meditations*, which says, "There are two things to aim at in life: the first, to get what you want, and after that, to enjoy it. Only the wisest of mankind achieve the second." I am not very wise, but I try to follow this advice.

· 12 ·

Research in Post-Nobel Years:
From Superacids to Superelectrophiles

The Nobel Prize frequently puts so many new responsibilities and pre-occupations on the shoulders of the winners that it drastically changes their lives and affects their ability (or desire) to continue active research. I have heard some express the opinion that having received their prize at a more advanced age was in a sense a blessing, because it did not hinder their research work throughout most of their career. This indeed may be the case, although Alfred Nobel's intention was to encourage and facili-tate research through his prizes. However, as I mentioned in Chapter 11, I was determined that the Prize should not affect my life, and certainly not my research, significantly. Now, two decades later, looking back on these post-Nobel years I feel that I mostly succeeded. These were among my best and in many ways most rewarding years of research. Helped by my dedicated younger colleagues and associates and by the close col-laboration with my life-long friend and colleague Surya Prakash, who took over all of the administrative responsibilities and burdens and runs our research institute as its director. I was able not only to continue my research but to extend it into new and challenging areas of significant practical importance.

A Life of Magic Chemistry: Autobiographical Reflections Including Post-Nobel Prize Years and the Methanol Economy, Second Updated Edition. George A. Olah with Thomas Mathew.
© 2015 John Wiley & Sons, Inc. Published 2015 by John Wiley & Sons, Inc.

A significant part of my previous research was based on the study of carbocationic systems using superacids and hydrocarbon chemistry. In a monograph with Prakash and Sommer (Wiley, 1985) in *Superacids* and its updated, extended new edition (Wiley, 2009) *Superacid Chemistry* with Prakash, Sommer, and Molnár as coauthors we discuss also our own work in the field. The acids we were able to use and explored turned out to be up to billions times stronger than previously recognized "strong" acids such as concentrated sulfuric acid. Acids are, in a general way, electron acceptors. Concerning protic (Brønsted) acids, in the condensed state there is no such thing as the unencumbered, naked proton (H^+). Having itself no electron, H^+ will always attach itself to any potential electron donors [whether n- (nonbonded), π-, or even σ-electron donors such as H_2]. In the case of Lewis acids (such as $AlCl_3$, BF_3), their electron deficiency is affected by the stabilizing effect of ligands (neighboring groups) attached to the electron-deficient central atoms.

There are basically two approaches to enhancing the acidity of protic acids in the condensed state. The first approach involves decreasing the encumbrance of the proton by decreasing the nucleophilicity of the system. In aqueous media the proton is attached to water, forming the hydronium (more aptly, hydroxonium) ion, H_3O^+ (in its hydrated forms). In anhydrous HF it is H_2F^+, in fluorosulfuric acid, $FSO_3H_2^+$, and so forth. The leveling effect means that no system can exceed the acidity of its conjugate acid. Thus no aqueous acid system can have acidity exceeding that of the conjugate acid of water, H_3O^+. Because the proton transfer ability of H_2F^+ is higher than that of H_3O^+ or $FSO_3H_2^+$, acid systems based on HF are stronger acids than those based on oxygenated acids.

The second approach is the use of increasingly lower nucleophilicity anions and dispersion of the negative charge of the counter-anions (in the condensed state cations always must be balanced by anions). When anions are associated as, for example, via fluorine bridging going from SbF_6^-, to $Sb_2F_{11}^-$, $Sb_3F_{16}^-$, etc., charge is increasingly dispersed. In $Al_2Cl_7^-$, etc., chlorine bridging also occurs but is generally less predominant in bringing about association than fluorine bridging.

The vastly increased acidity of superacidic systems developed resulted in the significant new areas of superacid chemistry. Consequently, I began to ask myself whether a similar but more general approach could be used to produce electrophiles of greatly enhanced electron deficiency and thus reactivity. This resulted in developing the chemistry of super-electrophiles discussed in our book with Douglas Klumpp *Superelectrophiles and Their Chemistry* (Wiley, 2008).

Because of the leveling effect of the solvent (or excess acid itself acting as such) liquid phase acidity cannot exceed that of any conjugate of Brønsted or Lewis acid. In the case of water the limiting acidity is thus that of H_3O^+. Protonated water, H_3O^+ (hydronium ion), was first postulated in 1907, and its preeminent role in acid-catalyzed reactions in aqueous media was first realized in the acid–base theory of Brønsted and Lowry. Direct experimental evidence for the hydronium ion in solution and in the solid state came eventually from IR, Raman, and neutron diffraction studies. The gaseous ion was observed in mass spectrometric studies, including its clustered forms with up to some 20 water molecules, $H_3O^+ (H_2O)_n$ (Kabarle). This ion–neutral association (clustering) is somewhat like solvation in solution. In superacid solution, the hydronium ion was studied by ^1H and ^{17}O NMR spectroscopy. Eventually, hydronium ion salts with a variety of counter-ions such as SbF_6^-, AsF_6^- and BF_4^- were isolated and their x-ray structure was obtained (Christe). All of the experimental and high-level *ab initio* theoretical data support a pyramidal geometry for the hydronium ion.

protonated water (H_3O^+)   diprotonated water (H_4O^{2+})

Isotopomeric hydronium ions such as H_2DO^+ and HD_2O^+ were also prepared in $HSO_3F{:}SbF_5{-}D_2O/SO_2ClF$ solutions and characterized by ^1H and ^2H NMR spectroscopy, showing no exchange. Subsequently, however, Prakash and I, to our initial surprise, found that in the even stronger superacidic systems such as $HF/DF{:}SbF_5$, with increasing amounts of SbF_5 hydrogen/deuterium exchange takes place. Because no deprotonation equilibrium of the hydronium ion occurs in the extremely strong superacidic media and because H/D exchange occurs only with increasing acidity (i.e., with HF/DF:SbF$_5$ but not with HSO$_3$F/DSO$_3$F:SbF$_5$), the results suggested that the isotopic exchange takes place via diprotonated (deuteriated) isotopomeric H$_4$O^{2+} involved. However, in a limited equilibrium of mono- with diprotonated water, under extremely strongly acidic conditions the non-bonded electron pair of the oxygen atom of H$_3$O$^+$, despite being a positive ion, undergoes a second protonation.

$$HD_2O^+ \underset{}{\overset{H^+}{\rightleftharpoons}} H_2D_2O^{2+} \underset{}{\overset{-D^+}{\rightleftharpoons}} H_2DO^+ \underset{}{\overset{H^+}{\rightleftharpoons}} H_3DO^{2+} \underset{}{\overset{-D^+}{\rightleftharpoons}} H_3O^+$$

To study the structure and stability of the H$_4$O^{2+} dication, *ab initio* theoretical calculations have also been carried out. The H$_4$O^{2+} dication is isoelectronic with H$_4$N$^+$, CH$_4$, and BH$_4^-$ and has a tetrahedral minimum structure. Apparently, the double positive charge can be adequately accommodated in the tetrahedral structure to prevent spontaneous fragmentation. It should be mentioned that in H$_3$O$^+$ the oxygen atom is indeed highly negatively charged, the positive charge being essentially on the hydrogen atoms. Whereas thermodynamically H$_4$O^{2+} is unstable toward deprotonation to H$_3$O$^+$, it has significant kinetic barrier to deprotonation.

It must be recognized that the calculational data refer to idealized dilute gas phase and not to condensed state conditions, where solvation or clustering may have a stabilizing influence. In particular in small dications, solvation tends to diminish the charge–charge repulsion effect and thus in H$_4$O^{2+} bring into a thermodynamically more accessible region. The extra proton of H$_4$O^{2+} thus might be shared by more than one H$_3$O$^+$ in a dynamic fashion.

In subsequent studies, Schmidbaur succeeded in preparing and isolating the stable BF_4^- salt of $^{2+}O[AuP(Ph_3)]_4$ and determined its x-ray structure. Although, clearly, charge is significantly delocalized in the heavy metal ligand belt, nevertheless the isolation of an $O^{2+}L_4$ species is remarkable.

The significance of the possible diprotonation of water under extremely acidic conditions directly affects the question of acid strength obtainable in superacidic systems. The leveling effect mentioned above limits the acidity of any system to that of its conjugate acid. Thus, in aqueous systems, the acidity cannot exceed that of H_3O^+ present in equilibrium with H_2O.

$$H_2O + H^+ \rightleftharpoons H_3O^+$$

In superacidic systems, water is completely protonated and no equilibrium containing free water is indicated. However, the nonbonded electron pair of H_3O^+ makes it still a potential electron donor and can be further protonated at very high acidities (however limited the equilibrium with H_3O^+ may be). Thus, the acidity of such superacidic systems can exceed that of H_3O^+ and the "leveling out" is by that of H_4O^{2+}. We found that similar situations exist with other electrophiles, raising their electrophilic nature (electrophilicity) substantially.

$$H_2O + H^+ \rightleftharpoons H_3O^+ + H^+ \longleftarrow H_4O^{2+}$$

Alkyloxonium ions are derived from the parent hydronium ion, H_3O^+, by substituting one, two, or all three hydrogen atoms with alkyl groups (i.e., ROH_2^+, R_2OH^+, R_3O^+). Meerwein prepared salts of the trimethyl and triethyl oxonium ions, R_3O^+, and found them to be excellent alkylating agents for various heteroatom nucleophiles but not for alkylating hydrocarbons. In conjunction with protic superacids, such as FSO_3H, CF_3SO_3H, or $FSO_3H:SbF_5$ (magic acid), however, we found that they readily alkylate aromatics. For example, benzene and toluene were methylated and ethylated with trimethyl- and triethyloxonium salts in the presence of superacids. The protolytic activation of trialkyloxonium ions points to protonation (protosolvation) of the nonbonded electron pair of their oxygen atom (similar to that in H_3O^+), which enhances the electrophilicity of the alkyl groups. Lewis acid complexation has a similar effect.

$$
\begin{array}{cc}
\underset{\substack{R^{\diagup}\underset{R}{\overset{|}{O}}^{2+}R}}{\overset{H}{|}} &
\underset{\substack{R^{\diagup}\underset{R}{\overset{|}{O}}^{\delta+}R}}{\overset{\delta-LA}{\uparrow}}
\end{array}
$$

Acid-catalyzed alkylation of aromatics with alcohols themselves is widely used. Whereas tertiary (and secondary) alcohols react with relative ease, following a carbocationic mechanism, alkylation with primary alcohols such as methyl alcohol is achieved only in case of more reactive aromatics and under forcing conditions. Superacidic activation, however, again allows ready methylation of aromatics.

Similar activation takes place in the carbonylation of dimethyl ether to methyl acetate in superacidic solution. Whereas acetic acid and acetates are made nearly exclusively using Wilkinson's rhodium catalyst, a sensitive system necessitating carefully controlled conditions, and use of large amounts of the expensive rhodium triphenylphosphine complex, ready superacid-catalyzed carbonylation of dimethyl ether has significant advantages in preparing acetic acid.

$$
CH_3OCH_3 + CO \xrightarrow{\text{"H}^+\text{"}} CH_3COOCH_3
$$
$$
\downarrow H_2O
$$
$$
CH_3CO_2H + CH_3OH
$$

Acyl cations are relatively weak electrophiles. This is easily understood, because their structure is of a predominantly linear carboxonium ion nature, with the neighboring oxygen atom delocalizing charge and limiting their carbocationic nature.

$$
\overset{+}{R-C}=\overset{..}{O}: \quad \longleftrightarrow \quad R-C\equiv\overset{+}{O}:
$$

In their reactions with suitable nucleophiles, such as π-aromatics or heteroatom donor nucleophiles, the readily polarizable linear acylium ions shift a π-electron pair to oxygen, bending the ions and developing an empty p-orbital at the carbocationic center. This enables the reaction with aromatics. The acetylation of benzene can be depicted as

The lack of reactivity of acyl cations such as the acetyl cation with deactivated aromatics or saturated hydrocarbons is therefore not unexpected. However, under conditions of superacidic activation the oxygen atom of acetyl cation is protosolvated, greatly decreasing neighboring oxygen participation, enhancing the carbocationic nature, and thus acylating ability. This explains, for example, why the acetyl cation in superacidic media reacts with isobutane via hydride transfer involving a protoacetyl dication-like species, but in weaker acidic or aprotic media there is no reaction.

Apart from Brønsted acid activation, the acetyl cation (and other acyl ions) can also be achieved by Lewis acids. Although the 1:1 $CH_3COX–AlX_3$ Friedel–Crafts complex is inactive for the isomerization of alkanes, a system with two (or more) equivalents of AlX_3 was found by Volpin to be extremely reactive, also bringing about other electrophilic reactions.

$$C_nH_{2n+2} \xrightarrow[\text{20°C, 30 min}]{\text{RCOX + 2 AlCl}_3} i\text{-C}_4H_{10} + i\text{-C}_5H_{12} +$$

$$i\text{-C}_4H_9COR + \left[\begin{array}{c} \diagdown \\ C-C \\ \diagup \end{array}\right]_m$$

Volpin called the $CH_3COCl\cdot2AlCl_3$ complex an aprotic superacid. The results indicate that the acetyl cation is activated by further O-complexation with a second molecule of AlX_3 (the equivalent of protonation or protosolvation).

$$CH_3CO^+ AlCl_4^- \xrightleftharpoons{AlCl_3} CH_3\overset{+}{-}C\overset{\overset{+}{\diagup}O^- AlCl_3}{\underset{AlCl_4^-}{}}$$

$$\begin{array}{c} R \\ \diagdown \\ R \diagup \end{array}\overset{+}{=}OR' \longleftrightarrow \begin{array}{c} R \\ \diagdown \overset{+}{} \\ R \diagup \end{array}-OR' \xrightleftharpoons{H^+} \begin{array}{c} R \\ \diagdown \overset{+}{} \\ R \diagup \end{array}\overset{+}{\underset{H}{-OR}}$$

The results of activation of acyl cations led to our study of other *carboxonium* ions. Carboxonium ions are highly stabilized compared to alkyl cations. As their name indicates, they have both carbocationic and oxonium ionic natures.

Although the latter predominates, strong neighboring oxygen participation (in Winstein's terms) delocalizes charge heavily onto oxygen and renders the ions substantially oxonium ions. Superacidic activation protonates (protosolvates) the oxygen atoms, thus decreasing their participation, and enhances carbocationic nature and reactivity.

Carboxonium ions, for example, do not react with alkanes. However, in superacid solution, acetaldehyde (or acetone), for example, readily reacts with isobutane involving diprotonated, highly reactive carbocationic species.

The acid-catalyzed isomerization of alkanes (of substantial practical interest in production of high-octane gasoline) is a consequence of the ready rearrangement of the involved alkyl cations (Chapter 10). Carbonyl compounds (aldehydes, ketones) contain only a mildly electron-deficient carbonyl carbon atom of the polarized carbonyl group, which is insufficient to bring about rearrangements. Under superacidic activation, as we found in our studies, the carbonyl oxygen atoms not only protonate but get further protosolvated, significantly decreasing neighboring oxygen participation and development of carbocationic character, which allows rearrangements to occur. An example is the rearrangement of pivalaldehyde to methyl isopropyl ketone.

Acetic acid and other carboxylic acids are protonated in superacids to form stable carboxonium ions at low temperatures. Cleavage to related acyl cations is observed (by NMR) upon raising the temperature of the solutions. In excess superacids, a diprotonation equilibrium, indicated by theoretical calculations, can play a role in the ionization process.

$$CH_3CO_2H \xrightleftharpoons{H^+} CH_3CO_2H_2{}^+ \xrightleftharpoons{H^+} CH_3CO_2H_3{}^{2+}$$

$$-H_3O^+ \downarrow H^+$$

$$CH_3CO^+ \xrightleftharpoons{-H^+} CH_3\overset{+}{C}=\overset{+}{O}H$$

Protonation of formic acid similarly leads, after the formation at low temperature of the parent carboxonium ion, to the formyl cation. The persistent formyl cation was observed by high-pressure NMR in 1997 (Horvath and Gladysz). An equilibrium with diprotonated carbon monoxide causing rapid exchange can be involved, which also explains the observed high reactivity of carbon monoxide in superacidic media. Not only aromatic but also saturated hydrocarbons (such as isoalkanes and adamantanes) can be readily formylated.

$$HCO_2H \xrightleftharpoons{H^+} HCO_2H_2^+ \xrightleftharpoons{H^+} HCO_2H_3^{2+}$$

$$\downarrow{\scriptstyle -H_3O^+ \mid H^+}$$

$$CO \xrightleftharpoons{-H^+} HCO^+ \xrightleftharpoons{-H^+} \overset{++}{HCOH}$$

$$H_2O + CO_2 \rightleftharpoons H_2CO_3$$

The equilibrium of carbon dioxide with water forming carbonic acid is long recognized and is of great significance. Although carbonates are among the most abundant minerals on earth, free carbonic acid is elusive in solution. Protonated carbonic acid in superacidic media is, however, remarkably stable. We observed it years ago (Chapter 7) by dissolving carbonates or hydrogen carbonates in cold $FSO_3H–SbF_5$ (or other superacids) when no CO_2 was evolved. The solutions decompose only at around 0°C to give hydronium ion and carbon dioxide. The SbF_6^- salt of $H_3CO_3^+$ could recently be isolated and its x-ray structure determined. The close analogy between protonated carbonic acid and the guanidinium $[C(NH_2)_3^+]$ ion explains its stability. Both are highly resonance stabilized. The guanidinium ion undergoes further protonation in superacid solution, and we observed the dication in such media by NMR spectroscopy. Similarly protonated carbonic acids can also readily undergo a second protonation to $H_4CO_3^{2+}$. The protolytic cleavage of carbonic acid, which leads to carbon dioxide, should proceed through protonated and diprotonated CO_2, that is, $O=C–OH^+$ and $HO–COH^{2+}$.

$$\begin{array}{c} CO_3^{2-}/ \\ HCO_3^- \end{array} \xrightarrow[-80°C]{FSO_3H–SbF_5/SO_2} C(OH)_3^+ \xrightarrow{-10 \to 0°C} CO_2 + H_3O^+$$

$$C(OH)_3^+ \xrightleftharpoons{H^+} (HO)_2\overset{++}{C}OH_2 \xrightarrow[-H_3O^+]{H^+} CO_2H_2^{2+}$$

$$\downarrow{\scriptstyle -H^+}$$

$$CO_2 \xleftarrow{H^+} CO_2H^+$$

Protonated and diprotonated carbonic acid and carbon dioxide may also have implications in biological carboxylation processes. Although behavior in highly acidic solvent systems cannot be extrapolated to *in vivo* conditions, related multidentate interactions at enzymatic sites are possible.

Similar to oxonium ions, our studies of *sulfonium* ions also showed protosolvolytic activation in superacids to give sulfur superelectrophiles. The parent sulfonium ion (H_3S^+), for example, gives H_4S^{2+} (diprotonated hydrogen sulfide) in superacids.

$$H_2S \xrightarrow{\ H^+\ } H_3S^+ \underset{\ }{\overset{H^+}{\rightleftharpoons}} H_4S^{2+}$$

Various sulfonium and carbosulfonium ions show remarkably enhanced reactivity upon superelectrophilic activation, similar to their oxygen analogs; so do selenonium and telluronium ions. The alkylating ability of their trialkyl salts, for example, is greatly increased by protosolvation.

Halonium ions, including hydrido or alkylhalonium ions, are similarly protolytically activated, indicative of protonation of the non-bonded electron pairs of their halogen atoms.

$$HBr \xrightarrow{\ H^+\ } H_2Br^+ \underset{\ }{\overset{H^+}{\rightleftharpoons}} H_3Br^{2+}$$

$$CH_3\overset{+}{X}CH_3 \underset{\ }{\overset{H^+}{\rightleftharpoons}} CH_3\underset{\underset{H}{|}}{X}CH_3{}^{2+}$$

Various other heteroatom-substituted carbocations were also found to be activated by superacids. α-Nitro and α-cyanocarbenium ions, $R_2C^+NO_2$ or R_2C^+CN, for example, undergo O- or N-protonation, respectively, to dicationic species, decreasing neighboring nitrogen participation, which greatly enhances the electrophilicity of their carbocationic centers.

$$R_2\overset{+}{C}-NO_2 \underset{\ }{\overset{H^+}{\rightleftharpoons}} R_2\overset{+}{C}-\overset{+}{N}O_2H$$

$$R_2\overset{+}{C}-CN \underset{\ }{\overset{H^+}{\rightleftharpoons}} R_2\overset{+}{C}-\overset{+}{C}NH$$

To explain the experimentally observed high reactivity of HCN and alkyl nitriles under superacidic condition, Shudo found that in the Gatterman and Houben–Hoesch reactions, diprotonated HCN (or nitriles) are involved as the *de facto* reagents ($HC^+N^+H_2$, $RC^+N^+H_2$).

$$RCN \; \underset{}{\overset{H^+}{\rightleftharpoons}} \; R-\overset{+}{C}=NH$$

$$\overset{+\;\;+}{RC}=NH_2 \; \underset{}{\overset{H^+}{\rightleftharpoons}} \; R-C\equiv\overset{+}{N}H$$

An efficient nitrating agent, which I introduced, starting in my early research in Hungary is nitronium fluoroborate $NO_2^+BF_4^-$ and related nitronium salts (Chapter 7). They are generally used in aprotic solvents (nitromethane, sulfolane, dichloromethane). Whereas a large variety of aromatics are nitrated with them in excellent yields, strongly deactivated aromatics such as *meta*-dinitrobenzene are not further nitrated under these conditions. In contrast, in superacidic FSO_3H or CF_3SO_3H solution, nitronium salts nitrate *meta*-dinitrobenzene to 1,3,5-trinitrobenzene in 70% yield as well as other deactivated nitrofluorobenzenes.

Similarly, whereas nitronium salts show no reactivity toward methane in aprotic media, nitration takes place, albeit in low yield, in FSO_3H to give nitromethane. These data indicated that the nitronium is activated by the superacidic media. Its highly increased reactivity is explained by the finding that the linear nitronium ion, despite being a positively charged species, undergoes protonation (protosolvation) to the protonitronium dication (NO_2H^{2+}), an extremely reactive superelectrophile. The linear nitronium $O=N^+=O$ itself has neither a vacant atomic orbital on nitrogen nor a low-lying molecular orbital. Its electrophilic nature is mainly due to its polarizability when interacting with π-donor aromatic nucleophiles. In nitrating aromatics, when NO_2^+ approaches the π-donor substrate the latter polarizes NO_2^+ by displacing an N–O bonding electron pair onto oxygen. This causes bending and the development of an empty p-orbital on nitrogen, allowing eventual formation of a nitrogen–carbon bond.

The finding that highly deactivated aromatics do not react with NO_2^+ salts is in accord with the finding that their greatly diminished π-donor ability does not suffice anymore to polarize NO_2^+. Similarly σ-donor hydrocarbons such as methane (CH_4) are not able to affect such polarization. Instead, the linear nitronium ion is activated by the superacid. Despite the fact that NO_2^+ is a small, triatomic cation, the nonbonded electron pairs on oxygen can further interact with the acid. This interaction can be considered to be protosolvation (the electrophilic equivalent of usual nucleophilic solvation), which tends to bend the linear NO_2^+ ion. In the limiting case, *de facto* protonation forms NO_2H^{2+}. The protionitronium (which has been observed by Helmut Schwarz by mass spectrometry) is a highly reactive "superelectrophile." Protolytic solvation, however, can suffice in itself to achieve activation without necessarily forming the free NO_2H^{2+} dication.

$$O=\overset{+}{N}=O \quad \underset{}{\overset{H^+}{\rightleftharpoons}} \quad \left[O=\overset{\overset{O}{\|}}{\underset{O}{N}}-OH \right]^{2+}$$

Whereas the proton (H^+) can be considered the ultimate Brønsted acid (having no electron), the helium dication (He^{2+}) is an even stronger, doubly electron-deficient electron acceptor. In a theoretical, calculational study, we found that the helionitronium trication (NO_2He^{3+}) has a minimum structure isoelectronic and isostructural with NO_2H^{2+}.

Charge–charge repulsion effects in protolytically activating charged electrophiles certainly play a significant role, which must be overcome. Despite these effects multidentate protolytic interactions with superacids can take place, increasing the electrophilic nature of varied reagents.

A way to overcome charge–charge repulsion when activating the nitronium ion was found in our work when Lewis acids were used instead of strong Brønsted acids. The Friedel–Crafts nitration of deactivated aromatics and some aliphatic hydrocarbons was efficiently carried out with the $NO_2Cl.3AlCl_3$ system. In this case, the nitronium ion is coordinated to $AlCl_3$.

$$\left[O=N-O \longrightarrow AlCl_3 \right]^+ \quad Al_2Cl_7^-$$

The protolytic activation of hydrocarbon ions (carbocations) can also be achieved. We have found that the electrophilic nature of trivalent carbenium ion centers can be greatly enhanced by decreasing neighboring group participation not only by π- or n-donor ligands but also by hyperconjugatively interacting σ-donor alkyl groups. For example, the hyperconjugative effect of the methyl groups in the *tert*-butyl cation can be protosolvolytically decreased, resulting in the limiting case in a diposi-tive carbenium, carbonium dication.

$$(CH_3)_3C^+ + H^+ \longrightarrow (CH_3)_2^+\overset{+}{C}CH_4$$

This may be a factor while acid-catalyzed transformations (isomer-ization, alkylation) of saturated hydrocarbons proceed preferentially in excess strong acid media.

In my group's studies on superelectrophilic activation it was found that not only superacidic solutions but also solid acid systems can bring about such activation. Solid strong acids, possessing both Brønsted and Lewis acid sites, are of increasing significance. They range from sup-ported or intercalated systems to highly acidic perfluorinated resinsul-fonic acids (such as Nafion-H and its analogues) and to certain zeolites (such as HZSM-5) and acidic oxides. One of the major difficulties in characterizing solid acids is the accurate determination of their acidity. Frequently used methods are based on kinetic rather than thermody-namic measurements, which can give data on catalytic activity of the solid acids but not necessarily on their acidities.

To explain how solid acids such as Nafion-H or HZSM-5 can show remarkable catalytic activity in hydrocarbon transformations, the nature of activation at the acidic sites of such solid acids must be consid-ered. Nafion-H contains acidic -SO_3H groups in clustered pockets (Fig-ure 12.1). In the acidic zeolite HZSM-5 the active Brønsted and Lewis acid sites are in close proximity (~2.5 Å).

Brønsted and Lewis acid sites in zeolites

In these (and other) solid superacid catalyst systems, bi- or multiden-tate interactions are thus possible, forming highly reactive intermediates. This amounts to the solid-state equivalent of protosolvation resulting in superelectrophilic activation.

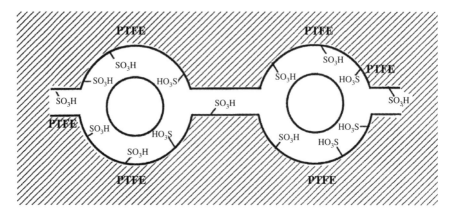

Figure 12.1. Sketch of a perfluorinated resinsulfonic acid similar to Nafion-H, showing clustering of SO_3H groups; PTFE = poly(tetrafluoroethylene).

Nature is able to perform its own transformations in ways that chemists have only begun to understand. At enzymatic sites many significant transformations take place that, in a generalized sense, are also acid catalyzed (e.g., electron-deficient metal ion-catalyzed processes). Because the unique geometry at enzyme sites can make multidentate interactions possible, these superelectrophilic chemical activations may have their equivalent in some enzymatic process. In recent years, Thauer, for example, found an active metal-free dehydrogenase enzyme. Because the involved mechanism cannot be that of metal coordination, we suggested it to involve carbocationic transformations (in a way again similar to superelectrophilic activation). Although I was never involved with enzymatic studies, sometimes in the literature, reference is made to this Olah-type enzyme.

My work on superelectrophiles emerged from my previous studies on superacidic carbocationic and onium systems. It led to the realization that a variety of electrophiles capable of further interaction (coordination) with strong Brønsted or Lewis acids can be greatly activated by them. Examples mentioned were onium and carboxonium ions, acyl cations, halonium, azonium, carbazonium ions, and even certain substituted carbocations and similar ones. This activation produces what I suggested should be called *superelectrophiles*, i.e., electrophiles of doubly electron-deficient (dipositive) nature whose reactivity significantly exceeds that of their parents. Superelectrophiles are the *de facto reactive intermediates* of many electrophilic reactions in superacidic systems (including those involving solid superacids) and should be differentiated from energetically lower-lying, much *more stable persistent intermediates*, which frequently are observable and even isolable but are not necessarily reactive enough without further activation.

Examples of some superelectrophiles so far studied and their parents are

$$
\begin{array}{cccc}
R-\overset{+}{\underset{R}{O}}-R & R-\overset{R_{2+}}{\underset{R}{O}}-R & R_2X^+ & R_3X^{2+} \\[2ex]
\overset{R}{\underset{R}{}}C=\overset{+}{O}R & \overset{R}{\underset{R}{}}\overset{+}{C}-\overset{+}{O}\overset{R}{\underset{R}{}} & C\overset{+}{X_3} & X_2\overset{+}{C}-\overset{+}{X}R \\[2ex]
RC\equiv\overset{+}{O} & R\overset{+}{C}=\overset{+}{O}R & R_2\overset{+}{C}NO_2 & R_2\overset{+}{C}-\overset{+}{N}O_2H \\[2ex]
R-\overset{OR}{\underset{OR}{C\overset{+}{}}} & R-\overset{+}{C}\overset{OR}{\underset{OR_2}{}} & R_2\overset{+}{C}CN & R_2\overset{+}{C}-\overset{+}{C}NH \\[2ex]
HCO^+ & H\overset{+}{C}=\overset{+}{O}H & R\overset{+}{C}=NH & R\overset{+}{C}-\overset{+}{N}H_3 \\[2ex]
HO-\overset{+}{\underset{OH}{C}}-OH & \overset{HO}{\underset{HO}{}}\overset{+}{C}-\overset{+}{O}H_2 & C^+(NH_2)_3 & (H_2N)_2\overset{+}{C}-\overset{+}{N}H_2 \\[2ex]
R_3S^+ & R_4S^{2+} & R_2\overset{+}{C}=NH_2 & R_2\overset{+}{C}-\overset{+}{N}H_3 \\[2ex]
R_3Se^+ & R_4Se^{2+} & H_2\overset{+}{N}N_2 & H_2\overset{+}{N}N_2H^{2+} \\[2ex]
R_3Te^+ & R_4Te^{2+} & NO_2^+ & NO_2H^{2+}
\end{array}
$$

It should be recognized that superelectrophilic reactions can also proceed with only "electrophilic assistance" (solvation, association) by the superacids without forming distinct dipositive intermediates. Protosolvolytic activation of electrophiles should always be considered in this context.

The new aspects of the broad field of onium ions including our work were discussed in my monograph *Onium Ions* with Kenneth Laali, Qi Wang, and Surya Prakash (Wiley, 1998). The emerging field of superelectrophiles and their chemistry is discussed in my book with Douglas Klumpp (Wiley, 2008) which includes Doug's studies during his postdoctoral stay and the work done subsequently on his own.

· 13 ·

Renewable Carbon Chemistry:
Anthropogenic Chemical Cycle with CO_2 Capture and Recycling to Methanol

The major effort of my work in the last two decades was taken up in close cooperation with Professor G. K. Surya Prakash, my former graduate student who became a close life-long friend, colleague, and my successor as Director of the Loker Hydrocarbon Research Institute. I was named as Founding Director besides being Distinguished Professor of Chemistry and Engineering working mainly to develop the practical, efficient, and environmentally neutral way to replace petroleum oil. With efforts and works of our devoted associates, we could develop the "Methanol Economy" concept (Chapter 15) into a reality with feasible practical applications of the underlying renewable carbon chemistry and the technology based on an anthropogenic (man-made) chemical carbon cycle. Our studies showed the way for renewable carbon fuels and derived economy making it suitable and feasible for the long-term benefit of mankind (*vide infra*).

Terrestrial life is inevitably bound to the essential element carbon through its compounds, which upon their oxidative biological or anthropogenic use eventually end up as carbon dioxide. Nature through its photosynthesis recycles carbon dioxide into plants and other derived

A Life of Magic Chemistry: Autobiographical Reflections Including Post-Nobel Prize Years and the Methanol Economy, Second Updated Edition. George A. Olah with Thomas Mathew.
© 2015 John Wiley & Sons, Inc. Published 2015 by John Wiley & Sons, Inc.

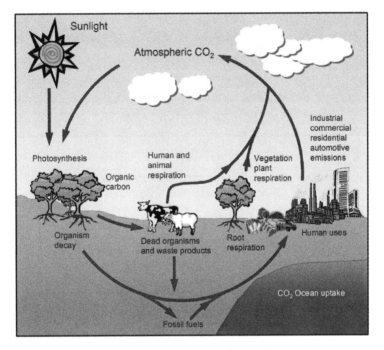

Figure 13.1. Natural carbon cycle.

systems and eventually to fossil fuels, which are used by mankind for energy generation and its derived products are essential for modern day life. I suggested in the 1990s and developed with my colleagues a chemistry-based new approach for the technological capture and recycling of carbon dioxide (CCR, i.e., carbon capture and recycling), thus supplementing nature's photosynthetic recycling and providing an inexhaustible source for carbon-based fuels and derived products. Whereas nature's own carbon cycle via photosynthesis (Figure 13.1) is an integral part of terrestrial life, natural replenishing of our diminishing fossil fuel resources would take hundreds of millions of years. Conversion of agricultural biomass or food sources (corn, sugar cane, etc.) to fuels and feedstocks is feasible and is already done. However, this starts to interfere with food production and prices and lacks the scale and sustainability required. Furthermore, growing energy crops require land, water, labor, energy, and also synthetic fertilizers (which are currently produced using fossil fuels and ammonia). The severity of our fuel and feedstock problem cannot be solved by the use of biofuels alone, which cannot satisfy the world's enormous requirements. A feasible solution to our carbon conundrum has been offered by our pioneering "Methanol Economy." In the last 15 years, my interest and research efforts based on my previous work centered on developing this new economic chemical approach to replace fossil fuels and produce derived hydrocarbon products. The concept expounds the use of methanol as a versatile fuel and feedstock, which can be prepared from varied carbon sources including recycling of carbon dioxide. This effort after a long concerted development is moving recently from laboratory research to increasing practical industrial interest and applications.

The "Methanol Economy" is based on the anthropogenic capture and chemical recycling of CO_2 to methanol and thus the utilization of carbon dioxide through methanol. This study was significantly escalated and became the center of focus of our Institute. It has been achieving greater worldwide attention not only as a new research and development sector but also for its scope into large-scale industrial application.

Development of feasible chemical methods to mitigate or eliminate the excessive carbon footprint of human activity and to provide a permanent inexhaustible supply of combustible and carbon neutral carbonaceous fuels and their products (which can be used without increasing the carbon dioxide content of the atmosphere) has been one of the major goals of our research efforts in this area. The Methanol Economy, which is based on the production of methanol fuels and chemicals by the capture and recycling of CO_2 could eventually replace the current fossil fuel (oil, natural gas, coal) based economy. By capturing excessive carbon dioxide that is released to the environment and subsequent recycling of the captured carbon dioxide via methanol via an anthropogenic carbon cycle to new carbon-based fuels and products allow future generations to continue to use carbon-based fuels and products from an inexhaustible source without causing harm to the environment. Thus, there is a need of an assured renewable source of these fuels and products without increasing the carbon footprint of our planet. Our new approach allows and makes possible the continued use of carbon-containing fuels and materials by rendering them renewable and environmentally adaptable.

The concept and development of the "Methanol Economy" presents significant advantages and great economic possibilities. Methanol serves also as an energy storage medium allowing not only convenient and safe storage and handling of bound hydrogen, but also transportation and dispensing of effective fuels for internal combustion engines and fuel cells. Derived dimethyl ether (DME) is an excellent diesel fuel and a replacement for natural gas and liquefied petroleum gas for household and industrial uses. Methanol and DME are also feedstocks for varied synthetic hydrocarbons such as ethylene and propylene and their products currently obtained from oil and natural gas. Use of nanomaterial-supported polyethylenimine CO_2 absorbents developed in our work enables the efficient capture and concentration of industrial and natural CO_2 sources (even that of the air) and their conversion into methanol and its chemical transformation to other useful chemicals and materials that are currently derived from petroleum and natural gas sources. Carbon dioxide from natural sources (such as accompanying natural gas, geothermal energy source, etc.) or industrial sources (fossil fuel burning power plants, varied materials such as cement and aluminum producing plants) can be isolated by separation or sorption on suitable absorbent material. Even the low CO_2 content of our atmosphere can be separated and recycled. Selective oxidation of methane available from various sources (still existing natural gas, methane hydrates deposited on seabeds, agricultural biomass and biowastes, etc.) and reductive recycling of CO_2 with hydrogen of water to methanol constitute the essentials of the anthropogenic chemical carbon cycle crucial for CCR.

As mentioned, new effective absorbent materials we developed are polyamine polymers deposited on a nanostructured high surface area supports. A particularly suitable polymer is polyethylenimine supported on nanostructured fumed silica or alumina. This supplements or replaces presently widely used liquid ethylenimine absorbent systems. Methanol is then produced from the desorbed carbon dioxide. Needed hydrogen is obtained from still available natural gas (methane) sources or generated by electrolysis of water using any available alternative energy sources. Methanol can be further processed to produce dimethyl ether or reacted over a bifunctional acidic–basic or zeolytic catalyst to give ethylene or propylene. Ethylene and propylene, besides being used to obtain widely used polymers, are also sources for synthetic hydrocarbons and derived chemical products.

$(-NHCH_2CH_2-)_x$

Polyethylenimine (PEI)
Linear

$(-NHCH_2CH_2)_x[-N-CH_2CH_2-]_y$
$\quad\quad\quad\quad\quad\quad\quad\quad|$
$\quad\quad\quad\quad\quad\quad\quad\quad CH_2$
$\quad\quad\quad\quad\quad\quad\quad\quad|$
$\quad\quad\quad\quad\quad\quad\quad\quad CH_2$
$\quad\quad\quad\quad\quad\quad\quad\quad|$
$\quad\quad\quad\quad\quad\quad\quad\quad NH_2$

Polyethylenimine (PEI)
Branched

Polyethylenimine $+ CO_2$ ⇌   Carbamate   ⇌   Bicarbonate

Polyethylenimine used for CO_2 capture sorbents and interaction of amine with CO_2

Merits of the suggested and developed "Methanol Economy" are discussed in my book with Goeppert and Prakash *Beyond Oil and Gas: The Methanol Economy* published by Wiley-VCH (first edition in 2006; an updated and enlarged edition in 2009; the third edition in preparation). It was also translated into Chinese, Japanese, Swedish, Hungarian, and Russian.

In the recycling processes, carbon dioxide capture and if needed temporary storage, that is, carbon dioxide capture and sequestration (CCS) is the initial step. CCS presently is a developing technology to reduce CO_2 emissions from new and existing coal- and gas-fired power plants and large industrial sources. The major steps in CCS are (a) capture of CO_2 from power plants or industrial processes, (b) transport of the captured and compressed CO_2 (usually in pipelines), and (c) underground injection and geologic sequestration (also referred to as storage) of the CO_2 into deep underground or under the sea rock formations. These formations are often a mile or more beneath the surface and consist of porous rock that holds the CO_2 under the impermeable, nonporous layers of rock that trap the CO_2 and prevent it from the rapid migration to the surface.

CCS can indeed reduce emissions from large stationary sources of CO_2, which include coal- and natural gas-fired power plants, as well as certain industrial ethanol, cement, natural gas, etc., processing plants. EPA's Greenhouse Gas Reporting Program (GHGRP) lists facilities that

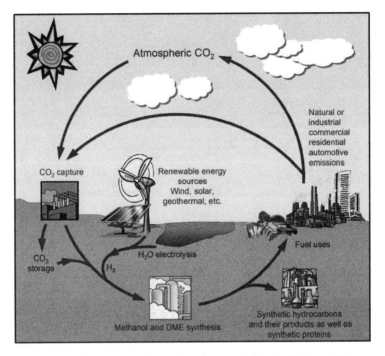

Figure 13.2. Anthropogenic chemical carbon cycle for the Methanol Economy.

capture CO_2 for the purpose of supplying the CO_2 to the economy or for injecting it underground. Currently CO_2 capture as well as extraction is occurring at over 120 facilities in the United States with about 70 MMt CO_2 (75% by capture and 25% by extraction), mainly on industrial processes, and the CO_2 is utilized for a wide range of end uses. End uses of CO_2 include enhanced oil recovery (EOR), food and beverage manufacturing, pulp and paper manufacturing, and metal fabrication. Figure 13.2 shows the portion of CO_2 that is currently being captured from power plants and other industrial facilities and the portion that is extracted by production wells from natural CO_2 bearing formations in the United States. CCS is however an essentially costly temporary storage of CO_2 which can diffuse back into the atmosphere or release at once in case of a geological event (earth quake, slides, etc.). It can be, however, followed up by chemical recycling of stored CO_2 to new fuels and materials (CCR) making CO_2 a valuable and renewable carbon source (*vide infra*).

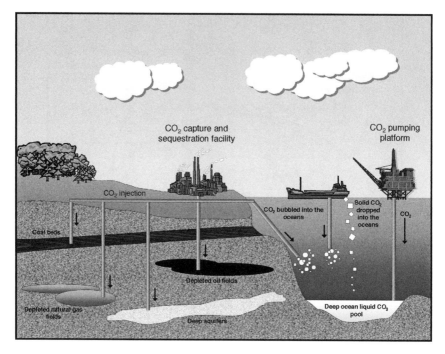

Figure 13.3. Overview of CO_2 sequestration technologies.

After capture, carbon dioxide (CO_2) is compressed and then transported to a site where it is injected underground for storage (also known as "sequestration"). CO_2 is commonly transported by pipeline, but it can also be transported by train, truck, or ship. Geologic formations suitable for sequestration include depleted oil and gas fields, deep coal seams, and saline formations (Figure 13.3). The U.S. Department of Energy estimates that anywhere from 1800 to 20,000 billion metric tons of CO_2 could be stored underground in the United States. That is equivalent to 600–6700 years of current level emissions from large stationary sources in the United States. Still CCS is costly and potentially temporary storage. As CCS becomes more widespread, it is expected that the portion of CO_2 captured in the United States from power generation and industrial processes will increase.

CCS technologies can capture up to 90% of CO_2 emissions from a power plant or industrial facility and store them in underground or sea geologic formations. Though carbon capture is an expensive technology, it has shown advance in power generation and other industrial processes. There are 12 active commercial-scale CCS projects at industrial facilities around the world (8 of those projects are in the United States), and approximately 50 additional projects are in various stages of development around the world. The world's first two commercial-scale CCS power plants, Southern Company's Kemper County Energy Facility in Mississippi and SaskPower's Boundary Dam Power Station in Saskatchewan, Canada, are under construction (expected to be completed in 2014). Interestingly, there is a growing market for utilizing captured CO_2 recently, primarily in enhanced oil recovery (CO_2-EOR). The International Energy Agency (IEA) estimates that CCS can achieve 14% of the global greenhouse gas emissions reductions needed by 2050 to limit global warming to 2°C. As already shown, CCS can allow fossil fuels, such as coal and natural gas, to remain part of our energy mix, by limiting the emissions from their use.

Potential sequestration sites must undergo appropriate site characterization to ensure that the site can safely store CO_2. Regardless, sequestrated CO_2 can migrate to the surface (effected by geological events such as earthquake) and is thus only a temporary solution. After being transported to the sequestration site, the compressed CO_2 is injected deep underground into solid, but porous rock, such as sandstone, shale, dolomite, basalt, or deep coal seams. Suitable formations for CO_2 sequestration are located under one or more layers of cap rock, which trap the CO_2 and prevent or minimize upward migration. Instead of just sequestering CO_2, it subsequently as we demonstrated can be converted to methanol and thus recycled (CCR).

Methanol, CH_3OH, as discussed is the simplest alcohol with least carbon and highest hydrogen contents contained of any liquid fuel. It is a very convenient liquid fuel with a high octane number (109). Though the most common methanol feedstock is natural gas today, it can also be produced from coal, municipal waste, landfill gas, wood wastes, etc. Carbon dioxide is however a major source for most present methanol production technologies. Its capture from industrial processes and atmosphere and recycling to methanol are gaining momentum in recent years. In the future, with advanced technologies, CO_2 extraction even from the atmosphere and oceans will

Figure 13.4. The George Olah carbon dioxide to renewable methanol plant in Iceland.

become feasible; overall production costs are expected to decline significantly.

In 2012 in Iceland, Carbon Recycling International (CRI) built and started operation of the first commercial CO_2 to methanol plant, named the *George Olah Renewable Methanol Plant* (Figure 13.4). This plant is based on Iceland's relatively inexpensive and abundant geothermal energy and the CO_2 which accompanies the natural geothermal steam. Numerous large conventional syngas-based methanol plants with an annual capacity of over 1M tons were and are being built in different corners of the world including China, Azerbaijan (Caspian Sea), Iran, the Middle East, Europe, and the United States based on natural gas or coal. Ongoing efforts to capture and dispose CO_2 (CCS) are planned and to some degree already practiced also in the Norwegian sector of the North Seas, Algeria, etc. They are dealing with CO_2 from natural gas or emitted from industrial sources such as fossil fuel burning power plants in the United Kingdom, Germany, Holland, Denmark, etc. The difficulties of underground or undersea storage of CO_2 (CCS) are however manifold. At best, it is only a temporary storage solution because CO_2 is volatile and will slowly seep back into the atmosphere, but in the case of a geological event (earthquake, slide, etc.), the instantly released CO_2 can have catastrophic consequences (as happened in

Cameroon in the 1990s). CCR in contrast offers a permanent solution via CO_2 recycling to methanol that provides a carbon neutral use of carbon-containing fuels and their products. It should be mentioned that even the possibility of recycling of CO_2 from the air was demonstrated although not used commercially.

The anthropogenic carbon cycle (via carbon capture and chemical recycling) is similar to the Haber–Bosch's nitrogen cycle through ammonia although in the long range even of greater significance. Using CO_2 as the carbon source with the use of alternative energies offers a sustainable alternative to traditional fossil fuel use while at the same time also mitigating the deleterious harmful environmental effect of CO_2 emissions, which contributes to global climate change. The chemical recycling of CO_2 with minimal carbon footprint to fuels such as methanol also provides a convenient and safe way to store and transport energy generated by any alternative energy sources such as solar, geothermal, wind, hydro, nuclear, etc. Many areas of the world still lack suitable grids to transport electricity. Methanol, being a suitable liquid, is easily stored and transported. It is also a proven broad use transportation fuel or feedstock to produce essential carbon-containing products such as olefins (ethylene, propylene) replacing all fuels and products currently derived from petroleum oil. Methanol–gasoline mixtures are excellent fuels for internal combustion engines. Recent discoveries of modified diesel engines allow also the use of methanol as fuel. DME is an already widely used diesel fuel substitute as well as substitute for natural gas and liquefied petroleum gas. Indigenously developed or modified diesel engines with squad assistance allow direct use of methanol and are already introduced in maritime uses. Methanol is an excellent high-density liquid hydrogen carrier for use in direct oxidation methanol fuel cells (DMFC), a practical technology and device (see Chapter 14). Even single cell and varied proteins can be produced using methanol for animal and eventually human food production. Single cell proteins were already industrially produced for a while from methanol by the former ICI in England.

Varied process studies to produce methanol from CO_2 and H_2 were reported on a laboratory scale. The capital investment for a methanol plant using CO_2 recycling and bi-reforming is estimated to be substantially lower than that of a conventional syngas-based plant. The key factor for the large-scale use of such a process is the nearby availability of the source materials replacing oil imports. Worldwide, presently more than 25 billion tonnes of CO_2 related to human activities are released into the atmosphere every year. Large amounts of CO_2 can thus be obtained relatively easily from various exhaust sources such as from fossil fuel burning power plants and varied industrial plants from cement factories to aluminum production to fermentation plants. Also, large natural CO_2 sources, such as CO_2 accompanying natural gas and geothermal energy production, could be captured and recycled to avoid atmospheric release. Even the low concentration of CO_2 contained in the air can be eventually separated and chemically recycled on a practical scale to methanol and derived synthetic hydrocarbons and their products.

Clearly, CO_2 can be recycled from industrial emissions to produce methanol and derived synthetic hydrocarbons and their products. CO_2 accompanying natural gas, geothermal, and other natural sources are also to be used. The CO_2 content of these emissions is high and can be readily separated and captured. The average CO_2 content of air is low (0.04%) but contributes to climate change (global warming).

The relationship between the atmospheric CO_2 content and temperature was first studied by Arrhenius as early as 1895. The climate change can be evaluated only over longer time periods, but there is a clear relationship between the CO_2 content in the atmosphere and earth's surface temperature.

Composition of Air

Nitrogen	78%
Oxygen	20.90%
Argon	0.90%
Carbon dioxide	0.04%
Water	Few % (variable)
Methane, nitrogen oxides, ozone, etc.	Traces

Recycling excess CO_2 evolving from human activities into methanol and dimethyl ether and further developing and transforming it into useful fuels and synthetic hydrocarbons and products will thus not only help our diminishing fossil fuel sources but at the same time help to mitigate global warming caused by human-made greenhouse gases.

Producing electricity directly from varied fuels is also achieved in fuel cells via their catalytic electrochemical oxidation, primarily that of hydrogen (Chapter 14).

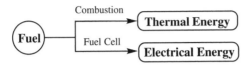

The principle of fuel cells was first recognized by William Grove during the early 1800s, but practical use was only recently developed. Most fuel cell technologies are still based on Grove's approach, with hydrogen and oxygen (air) combined in an electrochemical cell-like device, producing water and electricity. The process is clean, giving only water as the by-product. Hydrogen itself, however, must be produced in an energy-consuming process, using at present mainly fossil fuels and to a lesser extent the electrolysis of water. The handling of highly volatile hydrogen gas is not only technically difficult, but also dangerous. Nonetheless, the use of hydrogen-based fuel cells is gaining application in static installations in small electric devices (cell phones, laptops, etc.) and in specific cases such as space vehicles. Currently hydrogen gas is produced mainly from still-available fossil fuel sources using reformers, which convert them into a mixture of hydrogen and carbon monoxide from which hydrogen is then separated. Although this process mostly relies on our diminishing fossil fuel sources, electrolysis and other process to cleave water can provide hydrogen without any reliance on fossil fuels. Hydrogen-burning fuel cells, by necessity, are still limited in their applicability. U.S. Department of Energy (DOE) established the "Hydrogen and Fuel Cell Interagency Task Force," which includes representatives of 10 federal agencies for the coordination of efforts in research, development, and demonstration as well as federal adoption of hydrogen and fuel cell technologies. To promote further international cooperation, RD&D (Research, Development, and Demonstration), common codes and standards, and sharing of information, DOE works with the "International Partnership for Hydrogen and Fuel Cells in the Economy" (IPHE)—a partnership involving 17 countries and the European Commission. Through the International Energy Agency's (IEA's) two implementing agreements on hydrogen and fuel cells, the Department also coordinates the activities with more than 25 countries. Several major automakers, including Honda, Toyota, and Hyundai, are committed to producing commercial quantities of electric cars including hybrids utilizing fuel cells in the 2015–2017 timeframe and reaffirmed those commitments in 2012.

UKH2Mobility was launched in January 2012, bringing together three government agencies, the European Fuel Cells and Hydrogen Joint Undertaking, and 13 corporate participants (including major automakers such as Daimler, Toyota, Hyundai, Nissan, and Tata Motors), hydrogen suppliers (such as Air Liquide and Air Products), and fuel cell manufacturers such as ITM Power. In the second phase, UKH2Mobility develops a business case and strategy for the deployment proper of hydrogen infrastructure to support the anticipated introduction of FCEVs in the 2014–2015 timeframe. Anticipating the commercial availability of FCEVs in the 2015–2017 time frame, California Fuel Cell Partnership (CaFCP) also took significant steps in 2012 planning for more hydrogen fueling stations.

Despite all these efforts, the major safety challenges of preparation, safe transportation, and handling of hydrogen still remain. In the nuclear energy area, due to three major nuclear accidents in the last three decades, addition of new atomic energy plants and use of nuclear energy also are at stake. Many countries such as Austria, Belgium, Germany, Italy, Japan, Philippines, Sweden, and Switzerland started phasing them out. The modern world, however, will be totally paralyzed without ample sources of energy and power. With the fear of fast dwindling fossil fuels, increasing global warming concerns, alternate and nuclear (albeit rendered safer) sources are increasingly considered to satisfy world's demand for power. As innovative explorations for alternatives continue, there is no shortage of options which are explored and started to be used. For example, hydroelectric plants, wave farms, wind farms, parabolic solar energy collection farms, and geothermal energy production facilities are now already used all over the world to produce clean, safe, and sustainable energy. These forms of energy are also often cheaper, in the long term than nuclear power or any form of fossil

Figure 13.5. DMFC Cartridge and cartridge stack for laptop computers.

fuel with low or zero carbon footprints. The whole world now starts embracing these possibilities.

In our studies of fuel cells, we developed a new approach (discussed in Chapter 14) that uses directly convenient liquid methanol or its derivatives in fuel cells without first converting into hydrogen. The DMFC has been developed in a cooperative effort between our group at the University of Southern California and Caltech-Jet Propulsion Laboratory of NASA, who for a long time developed fuel cells for US space programs. The nearly hundred patents and pending patent applications cover direct liquid hydrocarbon fuel cells (including DMFC), electrode construction, membrane electrode assemblies (MEAs), fuel cell systems, methanol sensors and filters, water recovery, monopolar geometry, and new electrolysis method of methanol to form hydrogen.

Methanol fuel cells are already in commercial use now, in mobile phones, laptop computers, military equipment, and other applications being developed by electronics OEMs, such as Samsung and Toshiba, and other companies. Toshiba has developed efficient prototype stack of DMFC cartridges for laptop computers (Figure 13.5). Fuel cell powered notebook computers and mobile phones are projected to use approximately 100 disposable fuel cartridges during their workable lives and represent a recurring and potentially very large revenue stream. It has been projected that up to 22% of notebook computers and 2.5% of mobile phones will be powered by fuel cells in the future, which would result in demand for more than three billion cartridges per year. Methanol fuel cells and fuel cartridges are efficient, clean, and safe and have been approved to be carried on airplanes by the US Department

Figure 13.6. The Yamaha FC-Dii fuel cell scooter and Nissan fuel cell vehicle (FCV).

of Transportation and The International Civil Aviation Organization. These approvals are expected to encourage faster and more widespread adoption of fuel cells for portable electronic devices and fuel cell vehicles (Figure 13.6).

In DMFCs, methanol reacts with oxygen present in the air over a suitable metal or metal combination (Pt–Ru) catalyst, producing electricity while forming CO_2 and H_2O (Chapter 14).

Cathode reaction

$$1.5\ O_2 + 6H^+ + 6e^- \longrightarrow 3H_2O$$

Anode reaction

$$CH_3OH + H_2O \longrightarrow CO_2 + 6H^+ + 6e^-$$

Overall reaction

$$CH_3OH + 1.5\ O_2 \longrightarrow CO_2 + 2H_2O + \text{electrical energy}$$

More importantly, it was found that the process could be reversed. Methanol and the related oxygenates can be made from CO_2 via catalytic reduction without prior hydrolysis of water to produce hydrogen in what is termed a "regenerative fuel cell." This process can convert CO_2 and H_2O catalytically into oxygenated fuels (i.e., formic acid, formaldehyde, and methanol), depending on the electrode material and the potential used in the reverse fuel cell operation.

All combustion processes which use fossil fuels emit CO_2 as the end product, which however can be captured and recycled. In additional drilling operations, carbonate rock hydrolyte deposits in the ocean beds and many lakes around the world; CO_2 is present in enormous amounts making it an easily available base feedstock. Thus unlimited reserves and supply can be expected for the future which ultimately leads to a secure and stable economy. The reductive conversion of CO_2 to methanol will eventually be primarily carried out by catalytic hydrogenation using hydrogen produced by the hydrolysis of water (using any available energy sources such as atomic, solar, wind, and geothermal) or other means of cleavage (photolytic, enzymatic, etc.).

$$CO_2 + 3H_2 \xrightarrow{\text{catalyst}} CH_3OH + H_2O$$

Natural gas (in which methane is the major component), when available, can also be directly used for the CO_2 to methanol conversion, including improved energy efficient processes such as self-sufficient oxidative bi-reforming, which we developed recently. It uses partial methane combustion with oxygen of the air coupled with *bi-reforming* leading to the exclusive production of metgas (a mixture of H_2 and CO in 2:1 mole ratio) for subsequent methanol synthesis. The newly developed *oxidative bi-reforming* achieves direct oxygenation of methane to methanol in an overall economic and energetically efficient process leaving very little, if any, carbon footprint or by-products.

Effective direct oxidative production of methanol by direct methane oxidation is considered not feasible as it always results also in a mixture of thermodynamically as well as kinetically favored formation of further products that is, formaldehyde, formic acid, and eventually carbon dioxide. The practical synthesis of methanol by Fischer–Tropsch chemistry developed in Germany in the 1920s was based on syngas (a mixture of CO and H_2) which by separation and adjustment gave the needed 2:1 mixture. It was originally based on coal but subsequently shifted to more convenient natural gas. The conversion of natural gas using steam or dry (CO_2) reforming as well as partial oxidation produces H_2/CO mixtures with a molar ratio between 3 and 1. However, to produce methanol, a specific 2:1 H_2:CO mixture is needed, which we have named *metgas*.

Direct oxidation of methane

In conventional reforming processes, this ratio could not be reached effectively. When excess H_2 is present, CO_2 is added to the syngas mixture for the subsequent methanol synthesis, and the overall reforming process becomes strongly endothermic. In modern autothermal reforming processes (Holder-Topsoe, Lurgi, Linde, and others), part of the natural gas is combusted *in situ* to provide the needed heat energy. This renders, however, the needed *metgas* production even more complex and costly (up to some 60% of the overall cost of methanol plant) involving separation, purification, and adjustments such as water–gas shift reaction. We have succeeded to develop a facile new way called *bi-reforming* to produce *metgas* from methane (natural or shale gas) in a single step by combining the steam and dry reforming reactions by reacting methane or natural gas, CO_2, and steam in a 3:1:2 ratio over a Ni/MgO or related catalysts at 800–950°C in a pressurized tubular flow reactor at 5–30 atm.

Dry Reforming	$CH_4 + CO_2$ ⟶	$2CO + 2H_2$	$\Delta H_{298K} = +59.1$ kcal/mol
Steam Reforming	$2CH_4 + 2H_2O$ ⟶	$2CO + 6H_2$	$\Delta H_{298K} = +98.4$ kcal/mol

Bi-reforming $3CH_4 + 2H_2O + CO_2$ ⟶ $\boxed{4CO + 8H_2}$ $\Delta H_{298K} = +157.5$ kcal/mol

Metgas

Metgas by bi-reforming process, a combined steam and dry reforming

The overall *bi-reforming* process, being highly endothermic, requires substantial external energy as well as multistep feed preparation. As a significant further advance in our method, we found that an exclusive self-sufficient conversion of natural or shale gas (methane) to *metgas* for methanol synthesis can be achieved by complete *oxygenation* of part of methane using a process we call *oxidative bi-reforming*. Apart from the highly endergonic nature, traditionally practiced partial oxidation and reforming (autothermal reforming) necessitates costly separation steps and adjustments. The complete combustion of one equivalent of methane (natural or shale gas) with aerial oxygen gives the needed feed and process heat for the subsequent *bi-reforming* step. Thermochemical data shows that the exothermic reaction heat of methane combustion is more than sufficient for the subsequent endothermic *bi-reforming* process. It also provides the required CO_2–$2H_2O$ mixture, which is captured and admixed with 3 equivalent of methane (natural gas or shale) giving the needed specific feed for bi-reforming to produce exclusively *metgas* for the subsequent methanol synthesis step. This self-sufficient *oxidative bi-reforming* of methane to *metgas* can be carried out in a single bundled multitubular reactor or in separate reactors. The formed *metgas* is then converted to methanol in a well-known and industrially practiced synthesis step using $Cu/ZnO/Al_2O_3$ or related catalysts.

Thus, methanol produced from the still abundant natural or shale gas resources that can last well into next century can be used as a replacement for petroleum oil and its derived products while also decreasing their environmental harm (global warming) via CO_2 capture and recycling (CCR) in a feasible, economic way according to the framework of the "Methanol Economy" concept.

Methane Combustion

$$CH_4 + 2O_2 \longrightarrow CO_2 + 2H_2O \qquad \Delta H_{298K} = -191.9 \text{ kcal/mol}$$

Admixing of fresh methane followed by bi-reforming

$$3CH_4 + 2H_2O + CO_2 \longrightarrow \boxed{4CO + 8H_2} \qquad \Delta H_{298K} = +157.5 \text{ kcal/mol}$$
Metgas

Methanol Synthesis

$$\boxed{4CO + 8H_2} \longrightarrow 4CH_3OH \qquad \Delta H_{298K} = -86.8 \text{ kcal/mol}$$
Metgas

Overall Reaction

$$4CH_4 + 2O_2 \longrightarrow 4CH_3OH \qquad \Delta H_{298K} = -121.2 \text{ kcal/mol}$$

The self-sufficient oxidative bi-reforming and methanol synthesis

In the more distant future when methane sources will be exhausted, hydrogenative CO_2 recycling with hydrogen generated from water using solar or any other energy sources will provide an inexhaustible carbon source for methanol synthesis. Though the "Methanol Economy" is not yet widely used, it is regarded as economic and feasible and highly preferred over other suggestions as the "Hydrogen Economy" or "Bioethanol Economy."

Methanol clearly must be considered as a renewable economic *fuel of the future* and is started to be used in different countries. The "Methanol Economy" is thus becoming a feasible economic way to replace oil and eventually all fossil fuels. Its applications and advantages are described in our publications and monograph (*Beyond Oil and Gas: The Methanol Economy,* Wiley, 2009). Readers can refer to this as well as our some forty relevant patents based on our studies and inventions in the area.

· 14 ·

Rendering Methanol, A Renewable Carbon Source and Fuel:
Direct and Regenerative Methanol Fuel Cell

Another area of my post-Nobel research which turned into a major effort that continues now evolved from the realization that our fossil fuel-based hydrocarbon resources, the marvelous gift of nature in the form of petroleum oil, natural gas, and coal, are finite and not renewable. There is thus an essential need to find new ways to replace them. Our approach is based on anthropogenic carbon capture and recycling through renewable methanol, fuels, and derived products.

The rapidly growing world population, which was 1.6 billion at the beginning of the twentieth century, has now reached over 7.2 billion.

World Population (in Millions)

1650	1750	1800	1850	1900	1920	1952	2000	2014
545	728	906	1171	1608	1813	2409	6000	7200

Even if mankind increasingly exercises population control, by mid-century it can reach around 9.5–10 billion. This inevitably will put

A Life of Magic Chemistry: Autobiographical Reflections Including Post-Nobel Prize Years and the Methanol Economy, Second Updated Edition. George A. Olah with Thomas Mathew.
© 2015 John Wiley & Sons, Inc. Published 2015 by John Wiley & Sons, Inc.

enormous pressure on our resources, not the least on our energy resources to replace the diminishing fossil fuel sources. For its survival, mankind needs not only food, clean water, shelter, and clothing, but also energy. Since the cave man first managed to light fire, our early ancestors burned wood and subsequently other natural sources. The industrial revolution was fueled by coal, and the twentieth century added oil and gas and introduced atomic energy.

When fossil fuels such as coal, oil, or natural gas (i.e., hydrocarbons) are burned in power plants to generate electricity or to heat our homes and fuel our cars and airplanes, their hydrocarbons form carbon dioxide and water. Thus, they become nonrenewable (at least on the human timescale). We must find ways to eventually replace our diminishing natural resources such as fossil fuel in a renewable, economic, and environmentally adaptable, clean way. This represents a major challenge for humankind and chemistry.

With an increasing population, the world's per capita resources have difficulty to keeping up. Society's demands, however, must be satisfied while at the same time safeguarding the environment to allow future generations to continue to enjoy planet Earth as a hospitable home. Establishing an equilibrium between satisfying mankind's needs and safeguarding the environment is a further challenge which must be met.

Nature has given us a remarkable gift in the form of fossil fuels. However, what was created over the ages, man is using up rather rapidly. Since the industrial revolution, the use of coal became an essential part of our life. The subsequent large-scale use of petroleum and natural gas to generate energy and also as raw materials for diverse man-made materials and products (fuels, plastics, pharmaceuticals, chemicals, etc.) developed to full scale during the twentieth century. The US energy consumption (and that of the majority of the world) is indeed still very heavily based on fossil fuels. By 2010, energy from nuclear and other renewable sources (hydro, geothermal, solar) in the United States represented about 17%. While we witnessed significant developments in energy production from renewable sources, economic, environmental, and safety considerations (such as in the use of atomic energy) represent formidable challenges for the future.

US Energy Sources (%)

Power Source	Year				
	1960	1970	1990	2000	2010
Oil	48	46	41	39	37
Natural gas	26	26	24	23	25
Coal	19	19	23	22	21
Nuclear energy	3	5	8	8	9
Hydro, geothermal, wind, solar energy, etc.	4	4	4	8	8

Some industrial countries, despite all difficulties, increasingly get a significant part of their energy from nonfossil sources, such as hydro and atomic energy (vide infra). However, alternate energy sources such as hydroenergy or even solar energy (via various conversion processes) still represent only a small part of our overall energy picture. Whereas atomic energy is man's best hope for clean, practically unlimited energy for the foreseeable future, we must make it safer and solve problems of disposal and storage of radioactive byproduct waste. Influenced by catastrophic nuclear accidents, Three Mile Island (USA, 1979), Chernobyl (Russia, 1989), and the Fukushima (Japan, 2011), many countries even started phasing out nuclear power energy production. Therefore, a wide search for alternate clean energy sources such as wind, solar, and geothermal has become vital for the future sustenance of mankind.

Power Generated in Industrial Countries by Nonfossil Fuels (2010)

Nonfossil Fuel Power (%)

Country	Hydroenergy	Nuclear energy	Total
Canada	60	14	74
France	11	77	88
Germany	3.5	23.7	27.2
Japan	7.4	24.2	31.6
Italy	14	0	14
UK	1.4	13.8	15.2
USA	6.2	19.6	25.7

Oil use has grown to the point where the world's consumption is ~90 million barrels a day (a barrel equals 42 gallons, i.e., some 160 liters) or some 13 million metric tons. Fossil fuels are mixtures of varied hydro-carbons. As already mentioned, once we burn them, they are irreversibly used up and are not renewable on the human timescale. Fortunately, we still have significant worldwide fossil fuel reserves, including heavy oils, shale oil and gas, tar sands, and even larger deposits of coals (a complex mixture of carbon compounds more deficient in hydrogen), that can be utilized, albeit at a higher cost. I am not suggesting that our resources will run out in the near future, but it is clear that they will become scarcer, much more expensive, and will not last for very long. With the world population at 7.2 billion and still growing, the inevitable demand for oil and gas is particularly increasing although the past dire predictions of fast disappearing reserves were always overstated. Today our proven reserves with new discoveries and technologies are significantly higher than they were decades ago, but even more so is the future demand.

Recognized Oil and Natural Gas Reserves from 1960 to 2012 (in Billion Tons)

Year	Oil	Natural Gas
1960	43.0	15.3
1965	50.0	22.4
1970	77.7	33.3
1975	87.4	55.0
1980	90.6	60.8
1986	95.2	86.9
1987	121.2	91.4
1988	123.8	95.2
1989	136.8	96.2
1990	136.5	107.5
2002	160	160
2012	236	168.5

Proven oil reserves, instead of being depleted, as a matter of fact, tripled over the last 40 years reaching over 1.6 trillion barrels. Natural gas reserves have grown even more. This seems so impressive that many assume that there is no urgent oil or gas shortage in sight. However, inevitably increasing consumption by a growing world population and the drive for improving standards of living makes it more realistic to consider per capita reserves. Doing this, it becomes evident that our known reserves, if we go on using them as in the past, are limited. Even if we consider all other factors (new findings, savings, alternate sources, etc.) by the end of our century, we will face an acute major problem. Oil and gas will not be exhausted overnight, but market forces of supply and demand will start to drive prices up to levels that no one wants even to contemplate. By the end of the century, if we do not find new solutions, mankind will face a real crisis with grave economic and societal consequences.

All of mankind wants the advantages an increasingly industrialized society can give to its citizens. We all rely on energy, but the level of consumption is vastly different in the industrialized versus the developing world. For example, oil consumption per capita (bbl/day/1000 people) in China presently is only around 7, but it is more than 10-fold of that in the United States. China's oil use may double in a decade, whereas the bulk of its energy needs will continue to come from coal. This increase alone equals the amount of US consumption, reminding us of the size of the problem we are facing. This estimate does not consider that vast numbers of Chinese (or Indians, etc.), who want instead of riding bicycles, drive their own cars and use other energy-consuming conveniences at the level common in the industrial world. Many already enjoy varied technological advances of our time, necessitating increasing use of energy. Do we in the industrialized world have a monopoly on better life? I certainly do not think so.

Generating energy by burning fossil fuels including oil, gas, and coal at the present level, as I pointed out, will be possible only in the relatively short future, and even this causes serious economic and environmental problems (vide infra). The advent of the atomic energy age opened up for mankind wonderful new possibilities but also created dangers and concerns for safety. I believe, it is tragic that these considerations, however justified, already increasingly limit further development of atomic energy in many parts of the world. In fact, in the long run we have only limited alternatives to rely increasingly on atomic energy. We must solve, however, serious safety and environmental problems. Finding solutions is essential and possible if we have the will and determination for it. After all, in the United States, during World War II, we created the atom bomb by a great national effort. Could the same country that harnessed the energy of the atom in a wartime effort not also solve problems for its safe, peaceful use? I believe we can and will.

We burn our hydrocarbon resources to a great degree to generate energy and to use them as fuels. Diminishing supplies (and sharply increasing prices) will lead inevitably to the need to replace them by new technology. Synthetic oil or its hydrocarbon products will be, however, much costlier. Nature's petroleum oil and natural gas are still the greatest bargains we ever had. However, the price of a barrel of oil has already gone up above $100 in recent years (with market fluctuation). No synthetic oil manufacturing process is able to come even close to this price, and we will need to get used to this, not as a matter of any government policy, but as a fact of market forces.

Synthetic oil is feasible and can be produced from coal or natural gas via synthesis gas, a mixture of carbon monoxide and hydrogen obtained from incomplete combustion of coal or natural gas. Coal conversion to oil was used in Germany during World War II by coal hydrogenation or mainly through syngas $(CO–H_2)_x$-based Fischer–Tropsch synthesis. In South Africa during the 1960–1970 boycott years similar improved technology was used. The size of these operations, however, amounted to only about 0.5% of present US consumption. The processes further are highly energy-consuming, wasting up to half of the coal or natural gas used. They hardly can be the technology of the future. New and more economic processes are needed. Some of the needed new basic chemistry and technology are now developed. Such effort is also going on in our Loker Institute which is discussed subsequently.

We still have significant oil and natural gas resources. They are presently even expanded by new discoveries, for example, shale oil and gas and other new sources, around the world from polar regions to the depths of the oceans. Heavy oil sources and and tar sand are already increasingly utilized (Canada, Venezuela, etc.). Also we have vast deposits of methane hydrates (clathrates of methane with surrounding water molecules) under the tundras of Siberia and other polar regions as well as on the continental shelves of the oceans. Their extraction however represents formidable challenges. There is also the promise of finding large amounts of abiological "deep" methane formed not from biomass but by some abiological processes and are of carbonates or carbides.

We should also increasingly utilize the liquid hydrocarbons, which frequently accompany natural gas. These so-called "natural gas liquids" have presently primarily only heat value. They consist mainly of saturated straight hydrocarbon chains containing 3–6 carbon atoms, as well as some aromatics. Besides using them reforming, we found (see Chapter 8) that it is possible by superacidic catalytic treatment to directly upgrade these liquids to high-octane, commercially usable gasoline. Most of our natural gas (i.e., methane) is still burned to produce energy. However, their C_2–C_6 components are also used as sources for the synthesis of varied derived hydrocarbons, products, and materials. Nature provides us with the highest possible (4:1) hydrogen to carbon ratio in methane. The challenge is how we can best utilize it or natural gas to obtain liquid and higher hydrocarbons and derived products from it directly, without wastefully burning (albeit incomplete) it first to synthesis gas (a mixture of CO and H_2) to be used in Fischer–Tropsch chemistry.

Extensive studies were carried out to find ways for the selective oxidative conversion of methane or natural gas to higher hydrocarbons. Because combining two methane molecules to form ethane is itself endothermic by some 10 kcal/mol, oxidative removal of H_2 is needed to make the reaction thermodynamically feasible. Some selective oxidative

$$2CH_4 \xrightarrow{\ -H_2\ } C_2H_6 \qquad \Delta H = +10 \text{ kcal/mol}$$

coupling to ethane were reported, but only in low yields. In contrast, higher yield reactions give low selectivities. Superacidic oxidative condensation of methane to higher hydrocarbon was also explored in our work. More effective catalytic oxidative couplings are only a future promise.

A new approach we found during our studies was based on the initial bromination of methane to methyl bromide, which can be effected with good selectivity, although with relatively low yields. Methyl bromide is separated from excess methane, which is then recycled. Hydrolysis of methyl bromide to methyl alcohol and its dehydration to dimethyl ether are readily achieved. Importantly, HBr formed as a byproduct can be oxidatively recycled into bromine, making the overall process catalytic in bromine. We also succeeded in the direct conversion of methane to ethylene and higher hydrocarbons via the selective monohalogenation of methane.

$$CH_4 + Br_2 \xrightarrow[\text{cat.}]{} HBr + CH_3Br$$

$$\downarrow H_2O$$

$$CH_3OH + HBr$$

$$2HBr \xrightarrow{1/2O_2} H_2O + Br_2$$

$$\overline{2CH_4 + O_2 \longrightarrow 2CH_3OH + H_2O}$$

A commercial application in Brazil is developed by chemists from the University of California, Santa Barbara, and shows promise. An alternate application we studied is converting methyl bromide to ethylene.

$$2CH_3Br \xrightarrow{\text{catalyst}} CH_2 = CH_2 + 2HBr$$

This process eliminates the need of converting methyl bromide first to methanol which then can be converted to ethylene (or propylene). The conversion of methyl alcohol to ethylene using zeolitic catalysts (Mobil's HZSM-5 or UOP's varied zeolites) is commercially used on a large scale in China and some countries. Instead of zeolites, bifunctional acid–base catalytic chemistry, which we developed independently (using catalysts such as WO_3/Al_2O_3) can also be used.

$$2CH_3OH \longrightarrow CH_2 = CH_2 + 2H_2O$$

Our search for alternative fuels eventually resulted in using renewable methanol to make carbon fuels used for mass transportation and as a chemical source material for varied products from *metgas* that is, $2H_2$–CO produced from CO_2 capture and hydrogenative conversion.

Methanol, a convenient liquid fuel in its own right, we found can be made renewable by hydrogenative conversion of CO_2. Needed hydrogen can be produced from water by electrolysis or cleavage using any energy source primarily solar or other available alternate energy. This allows us to replace fossil fuels by an anthropogenic carbon cycle of CO_2 capture and conversion to *metgas*. This approach is entirely different from Fischer–Tropsch syngas chemistry based on fossil fuel resources. CO_2 chemical recycling via methanol, which also mitigates the global warming effect of excess of CO_2, and the concept of the "Methanol Economy" are discussed in the following chapter.

Burning of any hydrocarbon (fossil fuel) or, for that matter, any organic material converts its carbon content to carbon dioxide and its hydrogen to water. Because power plants and other industries emit large amounts of carbon dioxide, they contribute to the greenhouse warming effect on our planet, which causes significant environmental concerns. This was first indicated in 1898 in a paper by the Swedish chemist Svante Arrhenius (Nobel Prize, 1903). The warming trend of our earth can be evaluated only over longer time periods, but, from available data, there is a relationship between increasing CO_2 content of the atmosphere and the temperature (see Figures 14.1 and 14.2).

The increase of carbon dioxide in the atmosphere (even if its overall amount is only 0.04% or 400 ppm) affects our global climate, although other natural factors, not affected by human activity, also play a

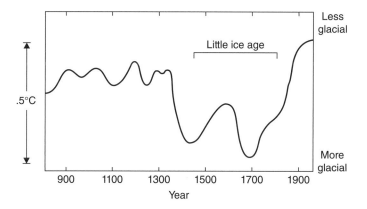

Figure 14.1. Estimate of the change in temperature in Europe over the past 1000 years.

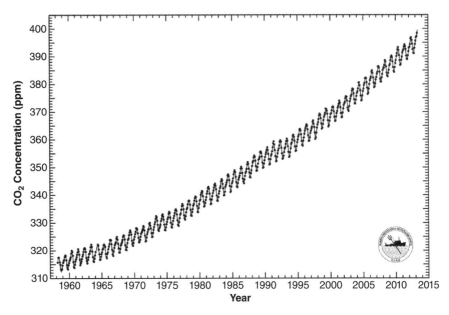

Figure 14.2. Concentration of atmospheric carbon dioxide–1958–2013 at Mauna Loa Observatory, Hawaii.

significant role in global warming and the potentially harmful effect of excess atmospheric carbon dioxide. Under public concern, in 1998 some 160 countries (over 190 countries by 2014) agreed in the Kyoto Protocol to limit and, in the industrial world, to decrease carbon emissions. Achieving this, however, is difficult and could cause great economic hardship.

As discussed, the goal of decreasing excess carbon dioxide emissions can be achieved by carbon capture and recycling (CCR) despite a growing world population and increasing energy needs. Fossil fuels can be replaced by CO_2 recycling via methanol using any source of energy including solar, wind, hydroelectric, and other alternate energy sources. Atomic energy is clean with respect to greenhouse gas emissions but represents serious problems of safety and radioactive waste disposal. Nevertheless, from what we know today, we will have no choice but to use nuclear fission (or fusion) energy in the twenty-first century on a large scale. We should, however, make it safer and solve the radioactive waste problem (either by improved disposal methods or by new ways of atomic reactor technology). The needed nuclear fuels (uranium, thorium) are themselves not unlimited, but breeder reactors or eventually even controlled fusion may give mankind its needed long-range energy freedom.

The control of excess carbon dioxide emission from burning fossil fuels in power plants or other industries as well as biological sources of human activity with nature's photosynthesis alone cannot be achieved but has been suggested to be possible by different ways, of which CCS, carbon capture and sequestration (i.e., collecting CO_2 and injecting it to the depth of the seas), is presently the only commercially considered possibility. Besides the obvious cost and technical difficulties, this would only temporarily store, not dispose of CO_2. Combining it with our developed CCR technology can solve this problem.

For my part, although I may be somewhat of a visionary for my time, I see a feasible solution to the problem by chemical recycling of excess carbon dioxide emissions into methyl alcohol and derived hydrocarbon products (see also Chapter 15).

In photosynthesis, nature recycles carbon dioxide and water, using the energy of sunlight, into carbohydrates and thus new plant life. The subsequent natural formation of fossil fuels from the biomass, however, takes a very long time. We cannot wait for this natural process to take its course and must find our own solution. The question I raised is, can we reverse the process and produce methyl alcohol and derived hydrocarbons by chemically recycling carbon dioxide and water? The answer is yes. Not only proper technological approaches, but a strong political approval and support are also essential. The average carbon dioxide content of the atmosphere, as mentioned, is very low ($\sim$0.04%) and, therefore, it is difficult to recover CO_2 from air economically although improved absorption and membrane separation techniques are developed by our Institute. One must admire nature even more because plants seem to be able to do this effortlessly. However, CO_2 can be readily captured and recovered technologically from the emissions of power plants burning carbonaceous fuels or from fermentation processes, calcination of limestone, or other industrial sources containing higher concentrations. To convert carbon dioxide subsequently to methyl alcohol and derived products chemically, hydrogen is needed. Water of the seas or moisture of the air is an unlimited source for hydrogen, but it must be split (generally by electrolysis), which necessitates much energy. The availability of safer and cleaner atomic energy, as well as alternative energy sources, eventually can provide this. Use of photovoltaic solar energy, for example, is a real possibility in suitable locations and is increasingly used. Energy of the wind, waves, and tides can also be used. For the present, however, because we still cannot store electricity efficiently, use of the excess capacity of our existing power plants (either burning fossil fuels or using atomic energy) in their off-peak periods represents a way, because they could produce hydrogen that then would be used to catalytically reduce CO_2 to methyl alcohol and derived hydrocarbons. This would allow reversible storage of electricity and at the same time recycle carbon dioxide, not only to mitigate global warming but also to provide an unlimited renewable source for hydrocarbons and their products when our fossil fuel reserves are increasingly depleted.

Besides chemical catalytic reduction of carbon dioxide with hydrogen to methanol, which is already commercialized in Iceland and started to be used industrially, we have explored a new approach to recycling carbon dioxide into methyl alcohol or related oxygenates via aqueous electrocatalytic reduction using a reversed (a regenerative) fuel cell.

We found significant practical use of methyl alcohol also in our studies on fuel cells. In cooperation with Caltech's Jet Propulsion Laboratory (JPL), we have developed an efficient new fuel cell that uses methyl alcohol directly to generate electricity without the need to first catalytically convert it to produce hydrogen. Its regenerative reversed application enables the conversion of produced CO_2 and water back to methanol.

The fuel cell is a device that converts chemical energy directly into electrical energy. The fuel cell was discovered in 1839 by William Grove, who found that the electrolysis of water in dilute sulfuric acid using platinum electrodes can be reversed. When water is electrolyzed, hydrogen and oxygen are formed. What Grove found was that if hydrogen and oxygen are combined in a cell over platinum electrodes in dilute sulfuric acid, water is formed and, simultaneously, electricity is produced. On the anode the fuel is oxidized, on the cathode oxygen is reduced. Grove's discovery, however, remained a curiosity. Only a century later work on the hydrogen–oxygen fuel cell was taken up again. Unsuccessful attempts were made to use cheaper metal catalysts. After World War II, alkalies, and later phosphoric acid, were used instead of sulfuric acid to ensure good contact between the gas, the electrolyte, and the solid electrode. These types of fuel cells started to receive limited application in static installations. With the advent of the space age, the need for improved fuel cells arose. For the Gemini and Apollo space programs, and later for the space shuttle, JPL developed and built oxygen–hydrogen fuel cells using pressurized liquefied gases. These cells worked well, but the handling of liquefied hydrogen and oxygen is not only cumbersome but also highly dangerous and can result in explosions. The need for developing a new generation, versatile, relatively lightweight and simple fuel cell for transportation and other uses, replacing inefficient heavy batteries, resulted in our cooperation in the 1990s with Sri Narayan (now at our Institute) and colleagues from JPL Caltech. The cell we developed is no longer based on hydrogen [which also can be produced from various hydrocarbon fuels (instead of water electrolysis) by a catalytic converter device (called reformer), that is, a small syngas-producing unit, with CO separated (and oxidized to CO_2) from H_2]. Instead, our fuel cell directly utilizes methanol, a convenient, liquid fuel, which is oxidized at the anode. The proton exchange membrane (PEM) initially used was of perfluorinated ionomer polymer (DuPont's Nafion), the electrodes being Pt-Ru. The membrane contains large amounts of absorbed water to conduct protons efficiently; thus liquid water is present. However, because not only protons but methanol itself can cross through the membrane, diffusion of methanol from the anode to the cathode became a major problem. To overcome the crossover we eventually developed an efficient new membrane, which eliminated the problem.

The schematic diagram of the liquid-feed direct methanol fuel cell (DMFC) is as shown in Figure 14.3. The involved chemical reactions are:

$$\text{Anode} \quad CH_3OH + H_2O \longrightarrow CO_2 + 6H^+ + 6e^-$$

$$\text{Cathode} \quad 1.5O_2 + 6H^+ + 6e^- \longrightarrow 3H_2O$$

$$\text{Overall} \quad CH_3OH + 1.5O_2 + H_2O \longrightarrow CO_2 + 3H_2O$$

Figure 14.3. Liquid-feed direct methanol fuel cell (DMFC).

In the overall fuel cell process, methanol is oxidized to produce carbon dioxide and water while producing electricity. Unlike processes that burn fossil fuels to liberate heat, the fuel cell converts chemical energy directly to electrical energy, and therefore it is not limited in efficiency by the so-called Carnot principle (according to which the efficiency of any reversible heat engine depends only on the temperature range through which it works rather than on the properties of the fuel). Because the working temperature cannot be increased excessively, heat engines have limited efficiencies. The efficiencies of power plants (using, e.g., a combination of gas and steam turbines) can be enhanced to 40%, but efficiencies of internal combustion engines are less than 20%. In contrast, the DMFC approaches 40% efficiency and this can improve further.

Because the DMFC does not use hydrogen as such, it is not only a greatly simplified system but also a safe and convenient one. Methanol is a water-soluble liquid and can be safely transported and dispensed by existing infrastructure (at gasoline stations). The high efficiency of the fuel cell will allow, instead of internal combustion gasoline or diesel engines, methanol also as the transportation *fuel of the future* in electric cars. Its use can be also important in electricity generation replacing the polluting and less efficient diesel generators in many parts of the world where the electric grid or reliable emergency electricity source does not yet exist. It is already used in smaller, portable applications such as cellular phones, computers and other electric devices, replacing batteries that have limited capacity and lifetime. In a battery, the chemicals accounting for the electricity-producing process are enclosed. Once they are used up, the battery must be recharged or replaced. In a fuel cell, external fuel is passed through continuously. Thus, they operate as long as external fuel such as methanol is introduced. We operated our fuel cell in the laboratory for periods in excess of 5000 h. It is now licensed by the German company SFC (Smart Fuel Cell) AG and is currently used commercially for smaller devices and power generators.

The DMFC works on the principle that when methyl alcohol is reacted with oxygen or air over a suitable metal catalyst, it forms CO_2 and H_2O while producing electricity. Similar to the concept of Grove's original fuel cell which is based on the reversal of the electrolysis of water, it is possible to reverse the methanol fuel cell process and to convert CO_2 and H_2O electrocatalytically back to oxygenates (formaldehyde, formic acid, methyl formate), eventually methyl alcohol, depending on the cell potential used. In this way, the aqueous electrocatalytic reduction of CO_2 is achieved. Problems, not the least that of overpotential, must be however further solved. In aqueous solution, carbon dioxide is first reduced to the CO_2^- radical, which is further reduced in the presence of water to $HCOO^• + HO^•$ and then to $HCOO^-$. Although the standard reduction potential differences for the various CO_2 reduction reactions are small, the large overpotentials observed on metals are most likely due to the formation of the radical anion intermediate. Improvements in electrocatalytic reduction of carbon dioxide continue to lower the overpotential on metals, with the goal of maintaining high current efficiencies and densities. In any case, the regenerative fuel cell process has substantial promise.

The concept of the reversed fuel cell, as shown schematically in Figure 14.4, consists of two parts. One is the discussed direct oxidation fuel cell. The other is an electrochemical cell consisting of a membrane electrode assembly where the anode is comprised of Pt/C (or related)

Figure 14.4. Regenerative Fuel Cell System.

catalysts and the cathode of various metal catalysts on carbon. The membrane used is a new proton-conducting PEM-type membrane we developed, which minimizes crossover.

A regenerative fuel cell system can also be a single electrochemical cell in which both the oxidation of fuels (i.e., production of electric power) and reduction of CO_2 (to obtain fuels) can be carried out by simply reversing the mode of operation (Figure 14.4).

The conventional electrochemical reduction of carbon dioxide gives formic acid as the major product, which can be obtained with a 90% current efficiency using, for example, indium, tin, or mercury cathodes. Being able to convert CO_2 initially to formates or formaldehyde is in itself significant. In our oxidation liquid feed fuel cell, varied oxygenates such as formaldehyde, formic acid and methylformate, dimethoxymethane, trimethoxymethane, trioxane, and dimethyl carbonate are all useful fuels.

The chemical or electrochemical recycling of carbon dioxide into methanol provides a renewable carbon base to supplement and eventually replace our diminishing natural hydrocarbon resources. Methanol (or dimethyl ether), as discussed, can also be readily converted into ethylene or, by further reaction, into propylene.

$$2CH_3OH \xrightarrow{-H_2O} CH_3OCH_3 \xrightarrow{cat.} CH_2{=}CH_2 + H_2O$$

$$CH_2{=}CH_2 + CH_3OCH_3 \xrightarrow{cat.} CH_2{=}CHCH_3 + CH_3OH$$

Ethylene (as well as propylene) produced from carbon dioxide subsequently allows ready preparation of the whole array of hydrocarbons as well as their derivatives and products that have become essential to our everyday life.

Whereas the nineteenth century relied mostly on coal for energy as well as derived chemical products, the twentieth century greatly supplemented this with petroleum and natural gas. With our limited fossil fuel resources, chemical recycling of carbon dioxide can increasingly play a role and provide a renewable base for our hydrocarbon needs. I emphasize that the chemical recycling of carbon dioxide to methanol of course necessitates much energy, which, however, can be obtained from atomic, solar, or other alternative sources when our fossil fuel sources diminish. My vision and summary of already achieved results point to a feasible economic and technological concept the "Methanol Economy" (Chapter 15) which I suggested first in the 1990s. Methanol has already got increasing significance as a clean burning fuel and chemical source material and started to be used in varied countries (such as China and Japan). Current developments are discussed in the following chapter.

The world's energy and material sources have difficulty in keeping up with our still rapidly growing population and increasing technological society. New and more efficient ways are needed to satisfy demands so that a reasonable standard of living can be provided for all. Society's demands must be satisfied while safeguarding the environment and allowing future generations to continue to enjoy planet Earth as a hospitable home. Establishing an equilibrium between satisfying mankind's needs and safeguarding and improving the environment is one of the major challenges of mankind. As a chemist, I find it rewarding to have been able to continue to contribute to this goal.

· 15 ·

The Concept and Development of the Methanol Economy

The French Nobel Laureate Albert Camus said in his remarks, upon accepting the Nobel Prize for Literature in 1957, that the prize propels the individual to "the center of a glaring light." The prize comes with many commitments, invitations for giving numerous lectures, connected with much travel and all the burdens having become a public figure and celebrity. Nobelists therefore, with few exceptions, are not able to do anymore significant work and contributions to their field. I not only read Camus' remarks but learned from them and having decided, I would not let it change my life and my life-long dedication to research and teaching. Many may feel that this is a selfish attitude and in some way, they may be right. On the other hand, I was 67 years old when I was awarded the Nobel Prize and I thought I earned the right to continue in my own ways. I never thought that at my present age of 87, I will be still here and allowed to follow an active life in chemistry. I am most grateful for it. This chapter covers my research based on work of practical interest and replacing petroleum oil and other fossil fuels with the feasible alternate material methanol. A brief summary of the related aspects mentioned in previous chapters is also added. I was fortunate to have been able to develop a new concept which I called the "Methanol Economy" (briefly mentioned also in Chapters 13 and 14) and much enabling chemistry for its feasible and economic application. Methanol

A Life of Magic Chemistry: Autobiographical Reflections Including Post-Nobel Prize Years and the Methanol Economy, Second Updated Edition. George A. Olah with Thomas Mathew.
© 2015 John Wiley & Sons, Inc. Published 2015 by John Wiley & Sons, Inc.

(methyl alcohol, CH_3OH) is a convenient liquid (bp 64.6°C) made originally from wood (wood alcohol) and for nearly a century from coal and subsequently from natural gas via syngas in a 1:2 CO/H_2 ratio (metgas) on a substantial scale. We succeeded to prepare it via an improved new way involving *bi-reforming* (a combined steam and CO_2 reforming process) or *oxidative bi-reforming* (a self-sufficient process burning part of natural or shale gas for the needed energy and using the $2H_2O–CO_2$ exhaust mixture of steam and carbon dioxide to form the needed 2:1 ratio of hydrogen and CO for the exclusive methanol synthesis). It can be applied practically for any carbon source, including the anthropogenic recycling of carbon dioxide (CO_2) supplementing or replacing nature's own biological photosynthetic carbon cycle. This is a feasible economic way to achieve humankind's independence from increasingly depleting oil reserves and allow everybody to continue to enjoy the benefit of carbon-based chemical fuels and products in a renewable and energetically friendly way.

Much effort and investment were put into alternative fuels primarily bioethanol. However, its production (from corn or sugarcane) interferes with essential food products and necessitates costly subsidy. Cellulosic ethanol is even more expensive and difficult to cleanly produce. Oil and natural gas, the main fossil fuels besides coal, are not only our major energy sources and fuels but also the raw materials for a great variety of derived hydrocarbon products. These range from gasoline and diesel oil to varied petrochemical and chemical products, including synthetic materials, plastics, and pharmaceuticals. However, fossil fuels what nature provided as a gift, formed over eons, are being used up rather rapidly. The expanding world population (now over 7.2 billion, projected toward 10 billion by the middle or end of twenty-first century), and the increasing standard of living and demands for energy in rapidly developing countries such as China and India, is further putting growing pressure on our diminishing fossil fuel resources and making them increasingly costly. Whereas our coal reserves may last for another two or three centuries, readily accessible oil and gas reserves— even considering new discoveries, improved technologies, savings, and unconventional resources (such as heavy oils, oil shale, tar sands, methane hydrates, coalbed methane, etc.)—may not last much beyond the twenty-first century. Once fossil fuels are combusted or used, they form CO_2, a harmful greenhouse gas contributing to global warming. Its excessive amount is not naturally renewable on the human timescale by nature's photosynthesis.

To satisfy humankind's ever-increasing needs, all feasible alternative and renewable energy sources must be considered and used. These include biomass, hydro and geothermal energy, the solar energy, wind, waves, tides of the seas, and others. Nuclear energy although made safer will also have to be further developed and utilized. Our efforts mainly focused on dealing with the challenges of how to best store, transport, and use energy in an environmentally friendly way, rather than dealing with the question of energy generation itself. Most of our energy sources are used to provide primarily heat and electricity and coming in some way from the sun. Electricity is generally a good way to transport energy over land with a suitable grid where it exists. It is however very difficult to store electricity on a large scale. Large batteries, for example, are still inefficient, expensive, and bulky. Electricity transfer over long distance is still difficult and wasteful and the construction of transmission lines across oceans is not possible with present technologies. Besides finding new energy sources, it is therefore necessary to develop new and efficient ways to store and distribute energy from whichever source it is derived.

One of the previously suggested and explored approaches is the use of hydrogen. Hydrogen can be generated by water electrolysis or cleavage using any available energy source primarily solar or from hydrocarbon sources. Hydrogen is considered as a clean fuel in the framework of the so-called "Hydrogen Economy." Hydrogen is clean in its combustion, producing only water, but the required energy generation is not clean, derived usually from fossil fuels or other sources with their attendant pollution effects. Although hydrogen has some desirable attributes for energy storage and as a fuel, its extreme volatility and explosive nature (and other issues) will need to be resolved for it to be used on a practical scale as an everyday energy source and fuel worldwide. Hydrogen being the lightest element has thus serious limitations in terms of its storage, transportation, and delivery of energy. The handling of volatile and potentially explosive hydrogen gas further necessitates special conditions, including cryogenic and high-pressure technologies, special techniques to minimize diffusion and leakage, as well as strict adherence to extensive safety precautions. All of this makes the commercial wide use of hydrogen difficult and very costly. In addition, the needed infrastructure must be developed at very high cost which seems at least presently, economically prohibitive. In addition, hydrogen itself in many applications cannot solve our continuing need for hydrocarbons and their products.

Today the transportation market is the major user of oil. Oil products such as gasoline, diesel, jet fuel, and derived products are produced globally on an enormous scale. They are easy and relatively safe to handle, to transport, and to distribute, mainly because of a vast dedicated infrastructure that exists. Consequently, any preferred alternative for fossil fuels should be compatible with existing infrastructure and uses.

Over the years starting in the 1990s, I suggested a new, viable alternative approach as of how to use more efficiently our still-available oil and natural gas resources and eventually to free humankind from its dependence on oil and other fossil fuels. This is based increasingly on the use of methanol or derived dimethyl ether (DME), as well as their varied derived hydrocarbon products, which forms the basis of my suggested "Methanol Economy." As mentioned, at present methanol is prepared almost exclusively from synthesis gas (syngas, a varying mixture of CO and H_2). It is obtained from incomplete combustion and steam reforming of fossil fuels, mainly natural gas and coal, involving in some degree carbon dioxide and processing the obtained H_2–CO mixture to a 2:1 ratio (metgas) needed for subsequent reductive synthesis of methanol. Methanol can also be prepared from biomass, including wood (it was originally called wood alcohol), agricultural byproducts, and municipal waste, but these play only a more limited role. Our extensive work has shown that its production is also possible by the conversion of methane or natural (shale) gas (and other carbon source material) with CO_2 recycling and steam using *bi-reforming*, ($3CH_4 + CO_2 + 2H_2O \rightarrow 8CO + 16H_2 \rightarrow 8CH_3OH$, See Chapter 14). It avoids the initial preparation of complex syngas. Instead of fossil carbon sources (coal, natural or shale gas or oil), methanol can also be produced by reductive hydrogenation of CO_2 from natural or industrial sources as well as recycling of exhausts from fossil fuel-burning power plants, cement plants, and other industrial sources and eventually from the atmosphere itself. The hydrogen required for these processes can be generated from still existing hydrocarbons or water electrolysis using any energy source. Energy can be stored in the form of safe and easily transportable methanol. The chemical recycling of CO_2 also helps to mitigate climate change caused by human activity (due to excessive burning and use of fossil fuels). The concept and the development of the "Methanol Economy" was reviewed in our monograph with Goeppert and Prakash (Wiley-VCH 2006, second updated edition 2009, third edition in preparation) and in our selected publications and patents.

Methanol is an excellent fuel in its own right, with an octane number >100, and it can be blended with gasoline (also with ethanol). Alternatively, even neat methanol can be used in today's internal combustion engines (ICEs) and with minimal modifications in modified spark-assisted diesel engines. It is a versatile source material for the production of varied products and materials now obtained from fossil fuels. Methanol as discussed can also be efficiently used to generate electricity in fuel cells. This was initially achieved by first catalytically reforming methanol into H_2 and CO. The H_2, after separation from CO, is then fed into a hydrogen fuel cell. However, methanol can also react directly with air in the direct methanol fuel cell (DMFC) without the need for reforming. DMFC greatly simplifies fuel cell technology, making it already commercially available to a wide range of applications such as portable electronic devices (e.g., cell phones, laptops), as well as for motor scooters and other vehicles (See Chapter 13). It can also be used for electricity generators in emergency backup systems and in areas of the world where electricity is still not available from a grid. DME, readily produced from methanol, is an excellent substitute for diesel fuel as well as household gas for cooking and heating.

Another potentially significant application of methanol is the conversion of natural gas (methane) into it for its storage and safe transportation when pipelines are not feasible or available. Today, liquefied natural gas (LNG) is transported across oceans under cryogenic and pressurized conditions by using supertankers (>200,000 tons). The LNG is then unloaded at port terminals and fed into pipelines to satisfy ever-increasing needs. However, LNG is potentially hazardous due to possible accidents or terrorist acts. A single supertanker exploding close to a densely populated area might have the devastating effect similar to a smaller atomic bomb. Whilst hoping that such a situation will never occur, realistically we must be prepared to find alternative, safer methods for transporting natural or shale gas. Natural gas can also be converted into liquid methanol close to the source. Natural gas is converted into liquid methanol presently using existing complex technologies, which need to be replaced by new and improved ones, such as *bi-reforming* or direct oxidative conversion without prior production of syngas (*vide supra*).

Besides its use as fuel and for energy storage and transportation, methanol and DME serve as convenient starting materials for basic chemicals such as formaldehyde, acetic acid, and a wide variety of other products including polymers, paints, adhesives, construction materials, synthetic chemicals, pharmaceuticals, and even single-cell proteins. Methanol and derived DME can also be conveniently converted via a single catalytic step into ethylene and/or propylene (the methanol-to-olefin, MTO process), which are already used industrially to provide essential building blocks for the production of synthetic hydrocarbons and their compounds. Practically, all the hydrocarbon fuels and products currently obtained from fossil fuels by the petrochemical industries can be efficiently obtained from methanol, which in turn, as emphasized, can be produced by the chemical recycling of natural or industrial CO_2 sources (Figure 15.1). Methanol, therefore, has the ability to liberate humankind from its dependence on oil and exhaustible other fossil fuels by allowing them and the derived fuels to be produced by technical hydrogenative recycling of CO_2 or by *bi-reforming/oxidative bi-reforming* supplementing and in many cases eliminating limitations of nature's photosynthetic cycle.

The concepts of the "Methanol Economy" we developed are starting in some countries to be commercially adapted as they have significant broad advantages and possibilities. Methanol, together with derived DME, as mentioned can be used as: (i) convenient energy storage media; (ii) readily transportable and dispensable fuels for ICE, as well as diesel or modified spark-assisted diesel engines, varied vehicles and household uses as well as fuel cells; and (iii) as feedstock for synthetic hydrocarbons

Figure 15.1 Chemical CO_2 recycling and the "Methanol Economy".

and their products, including varied fuels, polymers, and even proteins (for animal feed and eventually human consumption). The regenerative carbon source for methanol will eventually be the CO_2 contained in the air, which is available to all on earth and practically inexhaustible. The required energy for the regeneration of CO_2 into methanol can be obtained from any available energy source.

It should be emphasized that there is no preference for any particular energy source needed for the production of methanol. All energy sources, including renewable sources (solar, wind, hydro, tide, biomass, biofuels, geothermal, etc.) or atomic energy (with increasing safety), can be used in the most economically, safe, and environmentally acceptable ways. The ready conversion of methanol into synthetic hydrocarbons and their products will ensure that future generations will have unlimited and inexhaustible access to the essential hydrocarbon products and materials that today form an integral part of our life. At the same time, the "Methanol Economy," by recycling excess atmospheric CO_2, will mitigate or eliminate one of the major adverse effects on many forms of climate changes caused by human activity, that is, global warming by burning and using carbon-containing compounds.

We developed the concept of the "Methanol Economy" and enabling chemistry over the past two decades. The use of methanol as a fuel and gasoline additive is attracting increasing interest, particularly due to shortages and higher fossil fuel prices. When discussing "alcohol fuels," the public generally means ethanol and its availability from agricultural sources, including fermenting corn (in the United States), sugarcane (Brazil), or by bioconverting various wastes or other natural materials even cellulosic sources. This may be feasible in certain locations, but the production and the use of bioethanol has its own limitations. Although the use of bioethanol is currently feasible in some countries such as Brazil and even part of the United States (the corn belt) under certain conditions, it will, at best, satisfy only a relatively small part of our overall fuel requirements and have significantly adverse effect on food product availability and prices when produced from corn, wheat, sugarcane, and other food crops.

Although methanol and ethanol are chemically closely related (ethanol is the next higher homologue of methanol, CH_3CH_2OH versus CH_3OH), when the public consider "alcohols" as fuels they usually fail to realize the significant differences between the two. The fermentation of agricultural or natural products is used primarily to produce ethyl alcohol. It can also be produced from other natural sources. Methanol was originally produced from wood and called wood alcohol. It is now produced nearly exclusively by synthetic processes. Industrial synthetic ethanol, which can be readily produced by the hydration of ethylene derived from natural gas is rarely mentioned. As noted, a major industrial source for the production of ethylene is now methanol (MTO process). Bioethanol prepared by the fermentation of natural sources is a renewable, nonfossil fuel-based fuel, but vast areas of land are required to grow the sugarcane, corn, wheat, or other crops from which it is produced. The agricultural production of ethanol is also highly energy demanding, and currently most of this energy comes from fossil fuels. A fundamental difference between the production of bioethanol and methanol is that the latter does not mainly rely on agricultural resources but can be readily produced from varied feedstocks, including still-available fossil fuels and significantly using chemical capture and recycling of carbon dioxide from natural or industrial sources making it reversible and environmentally benign.

Eventually methanol can be made from atmospheric CO_2, with hydrogen generated by the electrolysis of water, using any form of energy (solar, wind, atomic energy, and energy from other alternate renewable nonfossil fuel sources). In this way, humankind will be able to produce methanol by chemically recycling CO_2 from the air, which is accessible to all. The needed water is an inexhaustible resource of hydrogen on earth. In the interim, however, still-available fossil fuels can be efficiently and more cleanly converted into methanol using methods such as *bi-reforming*.

The Olah group has long been involved in the study of various new aspects of methanol chemistry. *Inter alia*, beginning in the 1970s, the selective oxidation of methane to methanol in superacidic media, preventing further oxidation, was studied. During the 1980s it was followed by the discovery of the bifunctional acid–base-catalyzed conversion of methanol or DME into ethylene and/or propylene, and through them to both gasoline-range aliphatic and aromatic hydrocarbons. These studies were conducted independently of the zeolite-catalyzed chemistry developed by Mobil (now ExxonMobil), UOP (now Honeywell), and others for the conversion of syngas-based methanol into hydrocarbons. Our recent studies including the *bi-reforming* or *oxidative bi-reforming* pave the way toward feasibly and economically new synthesis of methanol and derived products essential for daily life and applications.

Based on the general realization that our nonrenewable fossil fuel sources are diminishing, increasing efforts have been directed to finding solutions to counteract their depletion particularly as transportation fuels. More efficient and economic use has been explored together with increasing use of non-fossil fuel alternative sources and safer atomic energy. It should be emphasized, however, that in addition to finding new solutions to satisfy our overall energy needs, there will still be in the post-fossil fuel era a continued need for convenient and safe energy storage, transportation fuels, and for a multitude of synthetic hydrocarbon products that will necessitate the creation of vast quantities of synthetic hydrocarbons. Whilst bioethanol and the "Hydrogen Economy" *inter alia* as discussed cannot itself fulfill these needs, in appropriate places it can be useful. Hydrogen combined with the chemical recycling of carbon dioxide is a feasible alternative way to the proposed "Methanol Economy."

In summary, the goals of the "Methanol Economy" encompass:

◆ New and more efficient ways of producing methanol and/or derived DME from fossil fuel sources, primarily natural gas, by their oxidative conversion, without prior production of syngas. Even the long sought-after monooxygenation of methane to methanol without oxygenated by-product becomes possible with *oxidative bi-reforming* process as detailed in Chapter 15

$$CH_4 + 1/2\,O_2 \rightarrow CH_3OH$$

◆ Production of methanol by hydrogenative recycling of CO_2 from natural and industrial sources, and eventually from the air itself as an inexhaustible carbon source

$$CO_2 + 3H_2 \xrightarrow{\text{catalyst}} CH_3OH + H_2O$$

◆ The use of methanol and derived DME as convenient energy storage media for transportation fuels for both combustion engines and new generations of fuel cells, including DMFC

◆ The use of methanol as raw material for producing ethylene and/or propylene as the basis for synthetic hydrocarbons and their products

$$2CH_3OH \xrightarrow{-H_2O} CH_3OCH_3$$

$$\downarrow {-H_2O}$$

Synthetic hydrocarbons and derived products $\longleftarrow$ $H_2C{=}CH_2$ + or

$$H_2C{=}CH{-}CH_3$$

The new hydrocarbon chemistry we have developed has significant broad applications and flexibilities. It converts, for example, natural (shale) gas sources to liquid methanol and recycles CO_2 through methanol to fuels and its products. It also opens up effective, economic, and environmentally friendly ways to utilize diversely available carbon-containing natural resources including heavy oils, tar sands, and bitumen deposits. Carbon dioxide recycling represents a significant aspect of the proposed "Methanol Economy." Any available energy source (alternative energies such as solar, wind, geothermal, and atomic energy) can be used for the production of needed hydrogen and chemical conversion of CO_2. Improved new methods we developed for the efficient reductive conversion of CO_2 to methanol and/or DME include *bi-reforming, oxidative bi-reforming* with methane and their source materials, as well as ways of new catalytic or electrochemical conversions. Liquid methanol with a boiling point of 64.6 °C is greatly preferable to highly volatile and potentially explosive hydrogen for energy storage and transportation. Methanol and derived DME are excellent transportation and industrial fuels on their own for ICE, spark-assisted diesel engines, and household uses. Methanol and DME produced from methane (natural gas or shale gas) via *metgas* without going to conventional syngas or by recycling CO_2 can subsequently also be used to produce ethylene as well as propylene. These are the building blocks for the petrochemical industry based presently heavily on oil for the ready preparation of synthetic aliphatic and aromatic hydrocarbons and for the wide variety of derived products and materials, which are obtained presently from oil and increasingly from natural gas, on which we rely so much in our daily life. Carbon dioxide can be chemically transformed from a detrimental greenhouse gas adding to global warming into a valuable, renewable, and inexhaustible carbon source of the future allowing environmentally neutral use of carbon fuels and derived hydrocarbon products (Figure 15.2).

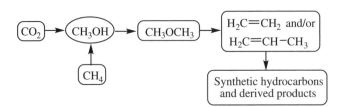

Figure 15.2. Natural gas (CH_4) and CO_2 as sources for methanol and derived products.

The "Methanol Economy" offers a feasible way to liberate humankind from its dependence on diminishing oil and natural gas resources while simultaneously utilizing and storing all sources of energies including both renewable and atomic. At the same time, by chemically recycling CO_2, one of the major man-made causes of climate change by global warming will be significantly mitigated while rendering carbon-containing fuels and materials regenerative.

Despite its high greenhouse effect and global warming tendency, due to the enormous amount and the easy availability of CO_2 in nature including large limestone deposits and other carbonates, it can be considered as powerful energy reservoir when we have effective ways of carbon recycling. On the ocean beds, in many lakes, reefs, and deposits large amount of CO_2 is buried underneath which can be eventually used. The "Methanol Economy" concept, therefore, represents a new approach of how humankind can decrease and eventually liberate itself from its dependence on diminishing fossil fuel reserves (coal, oil, and natural gas) while mitigating the global warming caused by excess carbon dioxide released by their combustion or other uses. It can be used also for the more efficient direct conversion of still-existing natural gas resources into methanol or DME, and on their production by chemical recycling of CO_2 from the exhaust gases of fossil fuel-burning power plants as well as the industrial and natural sources. Atmospheric CO_2 can also be captured and recycled using catalytic and electrochemical methods. This represents a regenerative chemical carbon cycle supplementary to nature's photosynthesis. It has developed into a reality of feasible practical solution to our carbon conundrum utilizing essential carbon-containing fuels to be produced in a renewable safe way for future generations when fossil fuels will be depleted. It should be emphasized that the "Methanol Economy" itself is not producing energy, but deals with the convenient storage and transportation of CO_2 for its conversion conveniently and safely to essential fuels and products. At the same time, it renders carbon-containing fuels renewable and environmentally neutral. This allows not only to mitigate the human-made cause adding to global warming but also to provide humanity with an inexhaustible, generally available, and renewable safe carbon source for ages to come.

· 16 ·

Gone My Way

Nearing to the end of my recollections of a full and rewarding career, I look back with satisfaction, but even more with gratitude. I had the good fortune that my career spanned the exciting time for science that was the second half of the twentieth and the early part of the twenty-first century, in which I was able to actively participate. I would like now to reflect on some of my beliefs and principles that always guided me. I cannot claim to have had great insight or to have wisely planned my career. My ways developed over the years mostly from my own experiences and circumstances, coming quite naturally. I am only too aware of my limitations and shortcomings. Thus, I do not want to give the impression that I consider "my ways" always to have been the right ones or even the most reasonable and certainly not the "safe" ones. We all are born with our inherited nature, but we are also affected by our environment, by our experiences, fortunes, and misfortunes, and, most of all by our free will and ability to overcome adversities and to succeed.

The fundamental education many Hungarian-born scientists received in our native country provided the foundation on which our subsequent scientific careers were built. At the same time, it must be remembered that most of us who became somewhat successful and noted did so only after we left Hungary. Denis Gabor (Nobel Prize in physics 1971 for holography), John Neumann (pioneer of computers and noted mathematician), John Kemeny (mathematician and developer of early computer programs), Leo Szilard (physicist), Edward Teller (atom

A Life of Magic Chemistry: Autobiographical Reflections Including Post-Nobel Prize Years and the Methanol Economy, Second Updated Edition. George A. Olah with Thomas Mathew.

physicist), Eugene Wigner (Nobel Prize-in physics 1963 for theory of atomic physics), George Hevesy (Nobel Prize in chemistry 1943 for isotopic tracers), like myself, all received recognition while working in the West. Only Albert Szent-Györgyi received his Nobel Prize (physiology or medicine 1937 for biological combustion processes with reference to vitamin C) while working in Hungary, but he also spent the last four decades of his career in the United States (Woods Hole Laboratory). Perhaps a small country could not offer its scientists opportunities for growth and work at home. Notably, no women scientist is on the list, perhaps telling us something about the still gender-separated educational system of the past and the bias against giving women an equal chance for a scientific career. I remember that in my university days there was not a single girl in our class! Hitler and Stalin also greatly contributed to the flight of scientist from Hungary. At the same time, I know that there were many highly talented contemporaries of ours who never left Hungary and somehow did not develop to their real potential. In any case, an isolated, small, and rather poor country, despite all its handicaps, turned out a remarkable number of noted scientists, mathematicians, engineers, composers, musicians, filmmakers, writers, economists, business leaders, so on.

I hope that working conditions and opportunities in Hungary will continue to improve. In 1989, Hungary rejected four decades of Communism (with the last decade of "goulash communism", only a milder left-over version) and started its development as a free, democratic country. It is not easy to achieve fundamental changes in a short time, but remarkable progress has been made. In 1984, when I first returned to Hungary after 27 years, the country was just beginning the changes that greatly accelerated after the turn to democracy in 1989. Hungarian science, which was treated relatively well by the Communist regime, was based substantially on a large framework of research institutes (run by the Academy of Sciences as well as state ministries and industries). At the same time universities were weakened, because much of the support for research was directed away from them toward these institutes. The process of reestablishing and strengthening university education and research has since progressed. The research institutes are reorganized into a more manageable scope and size, with combination of some and even privatization of others, at the same time establishing a close cooperation with universities and private industries. Of course, in the midst of fundamental economic changes and inevitable hardships, this is not an easy task. Many younger people, this time mainly for economic reasons and better opportunities, are looking to the West for opportunities. Whereas some experience in a foreign country is beneficial for the development of any scientist, it is hoped that Hungary increasingly will be able to offer opportunities for its young scientists that will induce them to stay or return home.

It is only natural that Hungarian-born scientists around the world who owe much to their heritage and the education of system of their native country are trying to help to their best ability to bring about these changes and improvements. For my part, I have in the past decades welcomed a series of younger Hungarian colleagues (primarily from the University of Szeged) for stays of up to 2 years in our Institute and we are maintaining continued contact and research cooperation with them after their return home. I also helped to endow a research award of the Hungarian Academy of Sciences (which generously named it after me) given in recognition of the achievement of younger chemists. Judy and I visited Hungary perhaps every other year till our health allowed it for lectures and miscellaneous events (the Nobel Prize itself has a great attraction in a small country). I always tried to emphasize to young people the fundamental need to get a good education and have faith in the future.

To the frequently asked question of what could I have achieved if I had not left in 1956, I have no definite answer. I believe, however, that we should look forward to the future and not second-guess of the past. It is the future that really counts and the key to the future (not only in Hungary but worldwide) is a good education. Only with a good education can the younger generation succeed, but of course circumstances, desire, and ability (as well as luck) are also significant.

Once I had decided on a career in chemistry, I was determined rather single-mindedly to make a success of it. I sometimes think about what would have happened had I chosen a different occupation or field. Having a rather competitive nature, I could probably have done reasonably well in a number of other areas. Certainly for some fields you must be born with a special talent. Musical talent, artistic ability, business acumen, leadership ability, and vision can be further developed, but there must be a given inner core. In creative science (and I emphasize "creative," not necessarily revolutionary in Thomas Kuhn's sense) there also must be some of this, but probably the acquired aspects are equally or more important. The genius of a Galileo, Newton, or Einstein is not questioned, but there are very few trailblazers who open up entirely new vistas. In mathematics, not unlike in music, you must have a talent born in you, and this is also the case in fields where mathematical thinking plays a major role. In contrast, in my field of chemistry, I believe it is not so much a specific inborn talent that matters but the ability for unconventional (perhaps it can be called creative) thinking and the ability to realize the significance when you come across something new. You also must have the ability and the staying power to subsequently explore it.

I strongly believe in the central role "multifaceted chemistry" plays in bridging other sciences and the ways it affects all aspects of our life. In a guest editorial for the Journal *Science* in December 1995 I wrote:

> "Humanity's drive to uncover the secrets of life processes and to use this knowledge to improve human existence has led to spectacular advances in the biological and health sciences. Chemistry richly contributes to these advances by helping to increase our understanding of processes at the molecular level, and it provides many of the methods and techniques of biotechnology. However, chemistry is not just an adjunct of biology and biotechnology. It is and always will be a central science in its own right."

I still believe what I wrote at the time. Chemists make compounds and strive to understand their reactions. Chemical synthesis, coupled with biotechnology, is well on its way to being able to reproduce many of nature's wonderful complex compounds and also to make unnatural ones. Although I started out as a natural product chemist coming from the Emil Fischer School, my own interests led me eventually to the chemistry of hydrocarbons. Hydrocarbons are essential to our everyday life because they make up petroleum oil, natural gas, and to a lesser degree even coal. Fossil (hydrocarbon) fuels generate energy and electricity, heat our houses, and propel our cars and airplanes. They are also the raw materials for most man-made substances, ranging from plastics to pharmaceuticals. What nature has given us over the eons, however, we are rapidly using up. As our nonrenewable fossil fuel reserves are diminishing rapidly and it will be up to chemists to synthesize hydrocarbons on an increasingly large scale and in new and economical ways. This thought also guided and is still guiding me, for example, in developing the "Methanol Economy" to replace oil, the fastest depleting fossil fuel and chemical raw material.

My longstanding research of hydrocarbon chemistry, particularly of acid-catalyzed reactions and their carbocationic intermediates, has also led to practical results. At Dow Chemical, I improved a series of processes even the production of styrene, a key product of the company. After my return to academia, we were able to improve other widely used acid-catalyzed industrial processes and to develop new ones. For example, in the production of high-octane and oxygenated gasoline, toxic and dangerous acids such as hydrofluoric acid are used in refineries. We found it possible to modify these by complexation with additives, mostly with pyridines, to make them substantially less volatile and thus much safer. New generations of highly acidic solid catalysts were developed for efficient and safe processes. So were environmentally adaptable and safe gasoline and diesel fuel additives, which allow cleaner burning and less pollution. To find ways to produce hydrocarbons by synthesis, as discussed in Chapter 13, we converted natural gas directly to liquid hydrocarbons and their oxygenated derivatives. Furthermore, I developed practically feasible methods to convert excess carbon dioxide even from the atmosphere into hydrocarbons with any source of energy (safer atomic and other alternate sources, including solar energy) needed to produce the necessary hydrogen. Using them can supplement nature's photosynthetic recycling of carbon dioxide by a technical carbon cycle and provide for our hydrocarbon needs after we have depleted our fossil fuels. Significantly, recycling carbon dioxide also mitigates global warming caused by excessive burning of fossil fuels or other hydrocarbon sources (See Chapters 14 and 15).

Chemistry, I fully realize, does not always enjoy the best of reputations. Many chemical plants and refineries are potentially dangerous and pollute their surroundings. At the same time, however, their products are needed, and our society enjoys a high standard of living that few are willing to give up. Our advanced way of life is in no small measure helped by the results of chemistry. Chemistry however should and I believe will be able to bring into equilibrium providing for our needs and responding to societal environmental concerns.

I have tried in my recollection to reflect on the search of my personal "magic chemistry" that led me through my career. Chemistry attracted me through the wide scope it offers and its many possibilities for exploring fascinating new areas. I enjoyed (and still do) the challenge and excitement to enter new areas from time to time and always found in them sufficient terra incognita for exploration of the unknown and unconventional.

The Indian-American physicist Subramanyan Chandrasekhar (Nobel Prize 1983) had a personal style of research, which I learnt, that seems to parallel mine. He intensely studied a selected subject for years. At the end of this period he generally summarized his work and thoughts in a book or a comprehensive review and then moved on to something else. He also refuted Huxley's claim "that scientists over 60 do more harm than good" by sharing the response of Rayleigh (who was 67 at the time) that this may be the case if they only undertake to criticize the work of younger men (women were not yet mentioned) but not when they stick to the things they are competent in. He manifested this belief in his seminal work on black holes, on which he published a fundamental book when he was 72 and his detailed analysis of Newton's famous *Principia* published when he was 84, shortly before his death. I have also written books and reviews whenever I felt that I had sufficiently explored a field in my research and it was time to move on. This indeed closely resembles Chandrasekhar's approach. I also always kept to areas in which I felt somewhat competent.

Much is said about at what age scientists become unproductive and should quit. As they grow older many cease doing research. Others just stay with their old fields or try to overreach in an attempt to produce something "lasting." They tackle problems, often with embarrassing results, outside their field of expertise. It is said that physicists do their best work in their 20s or 30s, whereas in other fields age is less of a detriment. I myself never felt influenced by such questions or considerations. May be I am lucky, but writing now at age 87 I still feel the drive and satisfaction of pursuing my search to understand and explore novelty in chemistry. If this goes away, I will indeed retire. There are many other satisfying things one can still pursue and enjoy besides research.

It has been my way to pursue my interests free of any a priori judgment of whether the topic was popular or well received at the time. Such considerations, at any rate, are temporary and tend to change with time. It is also more fun to be one of the leaders in a new area, regardless of its popularity, than just to follow the crowd or be part of a stampede in a "boom area" (which tends to end eventually). Of course, the price you pay when you go your own way is that not only is recognition of your efforts long in coming (if it comes at all) but also it is difficult to get support for your work. An inevitable price for the freedom of academic research is the necessity to raise support for it (in chemistry, mostly to support your graduate students and post-doctoral fellows). In our American system, government agencies such as the National Institutes of Health (NIH), National Science Foundation (NSF), Department of Energy (DOE), and various Department of Energy and Defense (DOE and DOD)-related research offices, that is, Army, Navy, Air Force, ARPA (Advanced Research Project Agency), are some of the major sources of support. These research agencies are basically independent from each other and are even in friendly competition. In my view, this helps to keep American science vibrant, highly competitive, and active. In most other countries, there is generally a single central research support agency. If somebody in that agency does not like your type of research, that is generally the end of it. In our system, although the competition for research support is fierce, the fact that there are a number of agencies who could be interested in supporting your work offers more flexibility. This does not mean, however, that you do not need to put a major effort into your research proposals. You also must have flexibility on occasion to adjust some directions of your planned work to match it up better with the interest of a specific agency. Proposals, as a rule, are strictly peer reviewed. One could disagree on whether this system always works out for the best, but I am a firm believer that, overall, it does succeed and nobody has come up with a better system yet. Of course, there are people who have a talent in writing proposals and show great "salesmanship." Others are less adept at it, although not necessarily lesser scientists. Funding my interests, private foundations and individuals also have their own difficulties.

It is particularly difficult for young people to break into the support raising system and for older ones to stay in it. Years ago, I suggested an approach to help the former. If someone gets a junior faculty appointment in a reputable academic institution, this should give assurance that he/she was sufficiently evaluated to justify the appointment (based on both background and also promise for independent research). Therefore, that young researcher should be trusted with an initial grant giving substantial freedom to pursue the chosen research. Most established researchers when they write proposals know part of the outcome based on work they have already done in the context of their existing funding. They are thus in a much stronger position to suggest future work with a high probability of success. In a way, under my suggested approach, we would give our younger colleagues an initial credit. At the end of the grant period, the work done would be reviewed as to whether it produced tangible results. If so, it would warrant extension of support (or denial if the results were insufficient). Needless to say, that my idea has not been successful, but I still believe it has merit. Nature, of course eventually controls the age problem of scientists but many are productive when others already give up. Concerning this end of the scale, it is difficult for someone of my age group to be entirely impartial. Aging, however, is not necessarily something to be measured entirely by calendar years. Some stay very productive and active for a long time, whereas others, for whatever reason, get bored or tired of research earlier but frequently cast along for years on safe, but not necessarily exciting extensions of their previous work. I have no easy solution to offer, but I believe that age *per se* should not be used to discriminate. There is, of course, the valid point that having only limited funds and research facilities it is necessary to open up possibilities for younger researchers, even if this means closing them off for some older ones. I do not want to question the validity of such consideration. What I believe is that ways could be found to act more tactfully, with some human consideration of elderly colleagues who still can make valuable contributions and compete well on merit with their younger colleagues.

I recall a personal experience concerning my research on carbocations and their related chemistry. Because of its significance and relevance to biological systems and processes, it was supported in part by NIH for nearly 30 years for which I am most grateful. In 1994, however, my support was not renewed. I was told that it was judged that no further significant or relevant new findings could be expected from my work. This happened at the time when I was making most interesting and significant research discoveries on the new concept of superelectrophilic activation (Chapter 12). As it turned out this has substantial general significance to biological systems and even enzymatic processes, among others. Coincidentally, shortly after I was turned down by NIH I won the Nobel Prize and received a congratulatory letter from the NIH Director saying how proud they are of supporting my work (in the present tense). Bureaucratic mix-ups of course happen, but at the time I was not amused. In the long run, however, things worked out well. The nonrenewal of my support reinvigorated my research effort. It is always rewarding instead of brooding over some adversity to find a way to overcome it. I was able to continue my work with continued support from other agencies as well as private supporters and foundations, who still had confidence in my work. I was able to prove that there was still much new pioneering chemistry left for me to discover and that the judgment of the futility of my proposed research was premature. As a matter of fact, I believe that my post-Nobel research turns out to be most exciting and productive.

Whereas we have eliminated most discrimination in our society, in the United States, age discrimination is a sensitive topic not frequently discussed. Officially, of course, it is not admitted, but de facto it very much exists. I have known many outstanding scientists, including Nobelists, who were forced to give up research at a time when they still felt that they could make significant contributions. Finding some ways to support them (which, on the overall scale of funding, would be not significant) is something worthwhile to consider. Some of our younger colleagues sitting in sometimes harsh judgment on review panels may not always realize that in years to come they may find themselves on the other side of the question, and I am sure by then their views will be different.

As mentioned, besides support by governmental agencies there is also support of research by industry, varied foundations and charitable organization, and individual donors. This all adds to the diversity of our research enterprise and its vitality. The US National Academy of Sciences itself does not support research to any significant degree. As a private organization it has no governmental budget or extensive private means. It, however, undertakes studies and reviews to advise the government and increasingly tackles significant, sometimes controversial questions of general interest on its own. This is different from some other countries, in which state-supported academies have their own research institutes. This was and to a degree still is the case in the former Eastern Bloc countries. In a different context, governmental support allows organizations such as the German Max Planck Society to maintain their numerous institutes for basic research or those like the Fraunhofer Society to maintain their more industrially oriented institutes. The closest we come in the United States to these institutes are the NIH Institutes and the National Laboratories, mainly concerned with health, energy, and military research.

With my European background, I was when I first came to America and still am impressed by the rather loosely organized, more decentralized way of research support. Of course, even in a great country like ours, resources are not limitless and inevitably prevailing trends of research set priorities. In my field of interest the 1970s and 1980s were a period when, after the two oil crises, research on catalysis, hydrocarbon fuels, and their synthetic preparation had significant public interest and support. Catalytic research in its many aspects was particularly much pursued and considered as one of the national priorities.

This changed to some degree in the 1990s. The unprecedented prosperity and economic upturn induced the public to believe that the concerns of our limited oil and gas reserves, of our energy needs in general are of no real urgency. The national science priorities heavily turned to the biological-life-health fields with strong emphasis on biotechnological developments. The development of electronics and computer-based technologies and industries, including development of new materials, were also spectacular and are continuing, but they are significantly funded by private industry and enterprise. I certainly do not want to minimize the enormous achievements and significance of these fields. It is obvious to me, however, that to maintain and build further our technology-driven development for the benefit of all mankind, we cannot neglect such basic questions as how we will provide the essentials for mankind's everyday life, including all the energy and materials we need. As a matter of fact, these because of increasing priority in recent years on our depleting natural fossil fuel resources put increasing emphasis on feasible ways to replace them or make them renewable. We also must maintain or restore a clean and safe environment for all the inhabitants of the Earth, who may reach 10 billion by the middle of the twenty-first century. To this end, it is clear that we must put more emphasis on research in areas such as those connected with energy, renewable resources for hydrocarbons, as well as solutions to environmental problems (as contrasted with only trying to regulate them).

I have discussed my efforts in some of these areas by starting to build a small university-based research institute on my move to Los Angeles in the late 1970s. I am proud that we have not only achieved this, but kept it going for nearly four decades. It is a privately supported Institute, with no major direct governmental (federal or state) funding. It is still a wonderful aspect of America that if you have a vision and work hard for it, you can achieve your goals. The Loker Institute raised the funds for its existence primarily from private donors, friends, and benefactors. We also established modest endowments to support pioneering work in new, high-risk areas and to provide the framework in which our faculty can compete successfully for traditional, competitive research support from governmental agencies for their specific research projects. Results of the Institute's work itself contribute to some degree to its operation costs through income from patents and intellectual property of our discoveries and other research efforts. The Institute also created a number of endowed chairs, giving our faculty further support and well-deserved recognition. We also have been able to endow our symposia and annual lectures in the field of our broad research interests, as well as maintaining our own essential library. Friends like the

Lokers, Moulton, and other supporters and foundations gave generously. The late Carl Franklin was particularly supportive in raising funds for the Institute, while himself was also one of our benefactors. I learned much from him, for example, that an essential aspect when you try to raise support is to contribute yourself to your best ability. It could be said more bluntly as "put your money where your mouth is." Nothing impresses others to contribute to your cause as much as your own example. Alfred Nobel may be also pleased that part of his prize money found its way to support our Institute. Judy and I are extremely grateful to our adopted country for the opportunities it gave us and also to the University of Southern California, which became an integral part of our life. Its presidents and board over the years always fully supported us. Returning some of our own good fortune was a privilege and a small expression of our gratitude.

I was fortunate that, during my career, I always had varied, broad interests and was able to explore a diverse palette of research topics cutting-across conventional lines of chemistry (for details of my selected published work with commentaries, see the three volumes with Prakash and Mathew of our book *Across Conventional Lines* by World Scientific Publishing Co., Singapore, 2003 and 2014). To follow my own interests and ideas came naturally. I always liked marching to my own drummer and was never scared to divert from the safe and well-traveled paths of "normal science" (in Thomas Kuhn's sense) to terra incognita (which many feel is unwise and most keep away from it). I indeed frequently disregarded what was the recognized limit of the terra firma of established knowledge. I reached "outside the box" to explore unknown areas and ideas of my own, despite the cautioning and even disapproval by some of my respected, more experienced and knowledgeable peers and friends. I had the good sense, however, to keep my eyes open and was prepared for the unexpected and unconventional (as well as to give up many unproductive ideas). When these came my way, I was able to recognize and follow them up. I also asked myself, when I found something interesting and new, whether it could be used for some practical use. Because I spent some years in industry and maintained an interest in the practical side, it was never difficult for me to freely cross the line between the basic and more applied aspects of my research.

When considering a new research project I had found it very worthwhile over the years to pause and ask what, assuming everything I was expecting to achieve would come through (and, of course, all researchers know that in reality this practically never happens), would be its significance. I recall Frank Westheimer, an outstanding Harvard chemist, recalling his encounter as a graduate student with James Conant, the eminent chemistry professor under whose guidance he started his work but who subsequently, upon becoming Harvard's president, moved away from chemistry. He inquired how his former student was doing in his research. Westheimer proudly told him what he had done and what he expected to do further. Conant politely listened and then said, "Frank, have you considered that even if you achieve all this, at best it may be just a footnote to a footnote in chemistry."

Indeed, it is very useful not only to consider how to achieve one's research goals, but also to ask early on the questions of its worthiness and significance when successfully carried out. It is the nature of research that once you are fully engaged in a project it becomes a very personal involvement. Being too close to it tends to influence your judgment. The question of whether the project, even if successful, would be worth the effort at this stage is usually not anymore considered. I believe this frequently leads to the proliferation of otherwise solid but not necessarily really significant or original research efforts. Even if you are blessed with a somewhat creative mind producing new research ideas, the most difficult part is to sort them out and decide which of them is really worth being followed up. This is also your responsibility when you become an advisor of your students or a research director in industry. The enthusiasm, hard work, and fresh ideas of your graduate students and postdoctoral fellows are of course the backbone of any research effort in your laboratory. It is, however, your responsibility to set the overall direction and to keep the work focused. In the course of any research project, unexpected and new observations can come along. You must keep your eyes open to recognize their significance, which can lead to new and frequently more significant directions than originally foreseen. The most difficult decision, however, is to know when to stop to pursue ideas that turn out to be unrealistic or unproductive. You must then redirect the work, learning from your disappointments and mistakes (some of which in retrospect you should have foreseen).

In some ways, this is also the case in industrial research, which I have also experienced. Projects frequently could be stretched out for a long time, even producing solid results, without, however, leading to real breakthroughs or solutions. As difficult as it is to admit that a research project does not lead to expected results or unexpected new findings, it is necessary if you indeed want to achieve may be something significant and new. At the same time, however, it is also essential to know how to stay the course and stick with your project and not give up prematurely because of some unavoidable disappointments, faulty starts, or other difficulties, without giving it your best shot as long as it is reasonable. I do not have a magic recipe for how to direct successful research. My experience, however, tells me that you should be able to find the right balance to make your decisions. It is also necessary to have faith to follow within reason your intuition and imagination, for which you eventually can be rewarded if some of your ideas indeed turn into successful reality. In any case you carry the full responsibility not only for success and recognition but also for failure.

As I mentioned, it is necessary to have a realistic degree of belief in your ideas and the ability to follow them through. If you yourself do not believe in your project, how can you convince your students and colleagues to give it their best effort? To build up their confidence is also essential. I always started my graduate students on some smaller project with a higher degree of probability for success. Once they gained confidence in their research ability they were ready to move on to higher risk, more challenging problems, without easily being discouraged by disappointments.

In my career, I have experienced many instances when my self-confidence was seriously tested. This is inevitable if you are not just pursuing regular, safe science. Because from my earliest days in research I preferred to go my own way searching the unexplored and unconventional, disappointments were unavoidable and frequent. When later on I inadvertently got involved in some controversies (most notably the nonclassical ion controversy), the general attitude of my colleagues and friends was not to take any chances and to "sit out" the controversy until overwhelming evidence led to its resolution. However, this was not my way. Once I was convinced of the validity of my relevant research approach, I was prepared to stay and stand up for it. The personal attacks and criticism which frequently came along were at times not easy to take. If you believe that your idea and work are correct (and of course, you must make sure through repeated, unequivocal work that your experimental results are correct), then you should be able to stand by your opinion, however lonely or unpopular it may be at the time.

Over the years as I mentioned, I followed the practice that, whenever in a specific area of my research I felt that I had substantially achieved my goals and that it was time to consider shifting my emphasis elsewhere, I wrote (or edited) a book or comprehensive review of the field. My books are listed in the appendix for interested readers who want to obtain more information or details (as well as to find relevant literature references).

My initial excursion into writing was in Hungary, where my university lectures on theoretical organic chemistry (really physical organic chemistry) were published in two volumes in 1955. A substantially revised German version of the first volume, written in 1956 with valuable input by Attila Pavlath, my Hungarian friend and colleague, was published in Berlin in 1960 (the second volume, however, was never realized due to circumstances). With few exceptions (as indicated), most of my subsequent books were published by my publisher John Wiley, Wiley-Interscience or Wiley-VCH). It started in the early 1960s when I met Eric Proskauer who at the time was running Interscience Publisher and Ed Immergut was his editor. Later, Interscience was bought out by John Wiley and we continued our relationship, with Ted Hoffman and others becoming my long-time editors and friends with whom it was always a pleasure to work. Wiley later also acquired the German publisher VCH and we continue a rewarding relationship. I also consulted for Wiley for many years concerning their organic chemistry publishing program and was able from time to time to suggest new authors and projects. I remember, for example, the start of Saul Patai's project, *The Chemistry of Fundamental Groups*, which later, with the involvement of Zvi Rappoport, Yitzhak Apeloig, and many others, developed into a truly monumental series. To keep the volumes from becoming obsolete, at one point, I suggested adding supplemental volumes to upgrade it and keep it current. These turned out to be very useful.

The Fiesers' *Organic Reagents* series, I remember, was originally Mel Newman's idea, but he himself was not interested in carrying it out. The Fiesers made it a real success. After Louis' death, Mary carried it on for many years, and now, at my suggestion Tse Lok Ho, a friend and former associate, continues it. Speaking of the Fiesers, they used to have a photograph of their cats adorning one of the front pages of their books. Being a dog lover, I felt that they deserved equal treatment and for nearly three decades the pictures of our Cocker Spaniels, Jimmy and Mookie, enlightened for an otherwise wasted front page of my books. With passing years, however, we gave up having further dogs. Our love and pride are now our grandchildren and it is only befitting that their picture adorns this book.

Two faithful friends, Jimmy 1969–1984 and Mookie 1985–1998

Writing or editing scientific books is always a rewarding experience, with the authors gaining most from the systematic study and review of the specific field. Even more important is to be able to evaluate and sort out what is really the important to be discussed and how to present it. It is also frequently a stimulation, because during reviewing a field, one frequently realizes important yet still unresolved problems or gaps worth studying or even gets ideas for new aspects to be explored. My books were always on topics to which I felt I made significant contributions. I was thus able to incorporate our own work and views and place them in proper perspective. I also hope that if future readers look up my books in libraries (assuming that printed books will survive and not be replaced entirely by electronic systems), they will find them still of some interest, use and even inspiration.

As rewarding as scientific publishing and writing is from an intellectual point of view, its monetary rewards are generally minimal (except for authors of successful college textbooks). I mention this without any regret or complaint. Royalties or honoraria never crossed my mind as a factor in any of my writings. Certainly a similar effort put into in any other endeavor would have been much more financially rewarding. There is however, the satisfaction to write something you believe will be of some value to your field and also the sheer joy in seeing your paper or book finally published, something that only those who have experienced it can fully appreciate. It is in a way like seeing your newborn child for the first time. Electronic publishing, however promising, I feel will not be able to give all the inspiration and incentive to authors to go through the long, demanding task of months or years to produce a comprehensive scientific review or book (or, for that matter, even research papers). Even if electronic computerized publications inevitably are taking a dominating role, I believe that printed journals and books at least in libraries should survive and provide continuous inspiration and wealth of knowledge for future generations.

Beside my research I always considered teaching an equally significant part of my professional life. As a parent (and later, grandparent), I believe that a good education is the single most significant asset we can give to our children. In a commencement address I gave at the University of Southern California in May 1995, I reflected on this.

> ... There is nothing more important and necessary in facing a dynamic new century than to be able to compete effectively in the arena of real life and to be able to offer the knowledge and skills demanded to succeed. We have undergone amazingly fast technological development in the relatively short period of two centuries since the industrial revolution. There is no single

aspect of our life which has not been touched and fundamentally changed by it. Progress is not only continuing but is accelerating. Just recall some of what happened in our own 20th century: the general use of electricity, dawn of the atomic age, fundamental changes in transportation (think about cars, planes, etc.) in communications (telephone, radio, television, satellite systems, FAX), the multitude of emerging new miracles of electronics, the enormous impact of the computer in all aspects of our life. We take all of them for granted as we near the end of the century. Whereas Science laid the foundations through developing fundamental knowledge, it was application by technology (engineering, manufacturing) which led to practical uses resulting in our highly technological oriented society. Of course, human knowledge and endeavor are broader in scope than just science or technology, and the arts and humanities much enrich our life. In life the most significant foundation for all of us is education and training which allow us to be prepared for a productive and rewarding life. Nobody will be able to compete effectively at any level of the work force without the education and skills needed in the 21st and following centuries, which is just around the corner. Technological advances will inevitably result in further increases in productivity, freeing more time for the individual to be able to spend in meaningful ways.

... Education is a pyramidal edifice—colleges and universities must build on the foundation laid by our primary and secondary schools. On their own, as fine as our institutions of higher education may be, they can only fail and be bogged down by flaws and shortcomings in earlier education. Schools further cannot do the job alone. It's the family which provides the surrounding and help that together with the schools give young people in their crucial years of growing up and maturing the environment essential to obtaining a solid education. One without the other cannot succeed. Society of course should and indeed try to help whenever the need arises, but neither Government nor private efforts will ever be able to replace the role and significance of the family...

I have always enjoyed teaching and direct contact with my students. Teaching chemistry varies in its approach at different levels. Early courses for nonmajoring undergraduates should provide sufficient, but not in-depth, introduction emphasizing the wonderful, magic world of chemistry and its broad applications as well as its significance to the other sciences. Regrettably, chemistry frequently is still thought of as a rather dry discipline based on physical laws. However, chemistry can be taught, even while acknowledging physical principles, as a vibrant, exciting topic with much relevance to and examples from our everyday life and the challenges and problems of mankind. Linus Pauling for example taught such a course at Caltech for years. There is indeed a return these days toward teaching more relevant interesting chemistry courses.

With the advance of recorded or "on-line" courses readily transmitted through the web or other electronic media, some argue that the need for direct, live teaching diminishes. This may be indeed the case for many courses including specific engineering or other professional updating or training ones. I believe, however, that nothing can replace direct personal contact between students and a teacher, who, in addition to information, also can give motivation and even inspiration. Furthermore, a good teacher should give much more than simply a recitation of a book, lecture notes, or a prerecorded one. If these were sufficient, there indeed would be no need for the direct classroom experience. However, attending lectures in which the topic is presented in a personal, dynamic, and, frequently, new way is a lasting experience. Furthermore, having direct contact with professors who, through their own knowledge and experience, can make the topic come alive for their audience is also a unique experience. Pauling's freshman lectures and Feynman's physics lectures at Caltech come again to my mind, as well as Woodward's Harvard lectures, in which he, in his precise manner, artistically built up on the blackboard edifices of organic synthesis. These are just a few examples of many remarkable science courses. If you actively participate in and contribute to science as well as are a dedicated teacher, you must continue to study and keep up with the field. Teaching and lecturing to your students and other audiences will give them much more than dry recitals. This is and should be the experience that students will remember long after they pass through your classroom.

At more advanced levels, chemistry is generally learned also by doing it. A young surgeon does not learn everything in the classroom or through anatomy bisections (these days, electronic simulation) or watching surgery being performed. Eventually he/she must step up to the operating table and, under the guidance and supervision of practiced peers, start to really learn surgery by performing it. The same is true for chemical research, which can be learned only by doing it. Laboratory education and practice prepares one, but nothing comes close to the real experience. I have never felt, therefore, that chemistry teaching and research can be separated. Not only are they inseparably mashed together, but the learning process never really ends. A research laboratory is not only a place to carry out research but also a challenging and inspiring environment that fosters continuous intellectual interaction and self-education. We never cease to learn, not only by keeping up with the scientific literature but also through discussions and interchange with our colleagues, labmates, and others. Graduate students and postdoctoral fellows, and even their faculty advisors, learn continuously from each other in an inspiring research environment, much more than any organized course or seminar can provide.

It is not necessarily the faculty itself (however good it may be) or the quality of the students (who frequently more than anything make their professors stand out) that makes some chemistry departments really outstanding. It is difficult to define the essential atmosphere that is conducive and collegial and at the same time permits competitive spirit that allows scholarly work and research to prosper. It is possible to build new laboratories or institutes, hire good faculty, and provide the prerequisites for their work, but eventual success will depend on whether an atmosphere can be developed that brings about the challenging and conducive spirit essential for innovative research and learning. Only this can help turn young, eager researchers into independent scholars capable of pursuing their own work.

An integral part during all my years of research and teaching effort has been a regular weekly meeting with my group (involving on occasions outside colleges). We discuss our research progress, on which individuals are asked rather randomly to report in an informal way. Questions are raised, and problems are debated. We also discuss some interesting new chemistry reported in the recent literature and occasionally have some more formal presentations (such as reviewing upcoming thesis defenses and presentation from the group at upcoming symposia, congresses, etc.). More recently the joint weekly meetings at the Loker Institute of the combined Olah–Prakash group have served to mutual benefit with Professor Prakash increasingly taking over leadership. They are very useful regular weekly events that also prepare our younger colleagues well for public presentations, not to mention job interviews.

I am fortunate in that early in my career I got used to lecturing and public speaking without written texts or even notes. For my courses I have prepared detailed outlines and collected and selected the material to be discussed, but this was only the framework, which I always changed and upgraded. This also assured of never getting bored while giving talks or lectures even on the same topic on different occasions, because I inevitably improvised and never really gave the same presentation twice. It also continuously brought in new aspects and left out others. Because my interests were (and still are) rather broad, I could lecture on a variety of topics, selecting those most appropriate for the specific occasion. I found it very useful to keep good contact with the audience, preferentially selecting someone for eye contact and, in a way, speaking directly to this person. In my classes, I also found it useful to involve the students through participation in an informal give-and-take fashion. Having a rather deep voice that came through rather well (of course with my unmistakable Hungarian accent) helped to keep my classes awake and to attract my audience's attention. If on occasion some still dosed off (a discouraging event for any speaker or teacher), I usually stopped and with some humor awakened them, promising to be more interesting and entertaining (which I tried to do). At the end of a lecture I knew myself best how I had fared, and the experience helped me to improve. Some of my colleagues spend much more time and concern preparing for their lectures, and some even agonize over them. I am, I guess, fortunate that I rather enjoyed lecturing and that it did not represent any particular stress for me. Of course, you still must prepare yourself and know your topic well, even more so than when you are using a prepared text. Answering questions and participating in discussions are particularly stimulating because you can gauge how our presentation and message really got through. You also can learn and get new ideas and stimulation from the comments and questions of your audience.

In chemistry, I think only in English (which is not always the case in other areas). When abroad, I lectured readily and spontaneously in German (although it took a few days to get back into the swing of it). Interestingly, it was somewhat difficult for me to lecture about chemistry in my native Hungarian (which we speak at home with Judy). Although I was 29 years old when I left Hungary and my mother tongue comes to me naturally, chemistry, as all the sciences, has its own dynamic changing language. After being away for nearly 30 years, when I first got back to Hungary and opted to give a lecture in Hungarian I found out how much the technical language had evolved (although eventually I managed better). I once spoke French rather fluently, but without practice it deteriorated to the point where I ecame hesitant to lecture in French (you can make mistakes in other languages, but in French it would be an offense.)

I have not kept records of the lectures I gave during my career at seminars, symposia, meetings, and congresses or at different universities around the world. There were many hundreds of these, and although these days I turn down invitations and do not travel outside, I still speak regularly in our research seminars. Over the years I visited and lectured widely across Europe and the World, and we have gone to annual trips to Europe. Since 1984 this also included visits to my native Hungary, generally arranged by a long-time friend Csaba Szantay. In 1995, we met with the President and Prime Minister of Hungary and I was honored with many Hungarian State Awards. I used these occasions to renew or establish contacts with friends and younger colleagues and also to visit with former members of the Olah-group scattered around the world. On a few occasions, I have visited Japan (my former postdoctoral fellows there number some 25 have an informal association and meet regularly) and Israel. We once visited India (from where I also had many associates) and enjoyed the hospitality of the family of Surya Prakash. I have also visited Honkong, Taiwan and later mainland China and found it vibrant and rapidly advancing. We have given up some years ago, with advancing age and health concern, travel but cherish memories.

In the mid-1960s with my "reluctant" sons on a European trip

In Hungary 1995 with President Göncz

With Prime Minister Gyula Horn

I still pursue science and direct jointly with Professor Prakash our group's work which takes up much of my time and energy. I particularly enjoyed over the years some visiting professorships or lectureships at different universities, which allowed close contact with faculty friends and students alike. I spent a semester at Ohio State in Columbus in the fall of 1963 as a guest of Mel Newman, who offered the use of his laboratory to Judy and me, and we did some rewarding work. I spent the spring of 1965 in Heidelberg as Heinz Staab and George Wittig's guest. In 1969, we spent a most pleasant month at the University of Colorado in Boulder hosted by Stan Cristol. In 1972, I was at the ETH in Zurich hosted by Heini Zollinger in his chemical technology department. In 1973, we were in Munich and had a particularly pleasant time with our friends Rolf and Trudl Huisgen, whom we visit regularly even now. Strasbourg in France was always one of our favorite cities, and in 1974 we spent a month at the university. Jean Sommer, as well as other faculty friends, always made this and other visits to Strasbourg wonderful events (including the culinary wonders of the region). We also visited and stayed on different occasions at the universities of Paris (Dubois), Montpellier (Commeyras, Corriu), Bordeaux, and Poitier (Jacquesy), among others, extending our experience in France. The University of London and particularly King's College represented, besides Durham (where I had close professional and personal ties with Dick Chambers and the late Ken Wade over the years), a special contact over the years. Since the late 1950, we visited London annually (where Judy's aunt Alice, with whom she was always very close, lived) and I regularly lectured at King's. The late Victor Gold, as well as Colin Reese and other friends, always extended great hospitality and collegiality, and as a honorary life lecturer of King's College, I always enjoyed my visits and talks. We also enjoyed stays in Italy, in Rome (with the late Gabriello Illuminati and Fulvio Cacace), in Padova (Giorgio Modena) and particularly in Milan with my late friend Massimo Simonetta and his wife Mirella. I had close interests and scientific cooperation with Massimo but first of all enjoyed the friendship of the Simonettas. Massimo also introduced me to the Italian chemical industry, and I consulted for some time and even served on the scientific board of ENICHEM.

I have always felt a special relationship with my students and post-doctoral fellows. Our joint research effort was only part of this relationship, which did not end with their leaving my laboratory. We usually keep in contact, like distant family members do, and see each other on occasions. I am proud of my "scientific family," and I hope they also have fond memories of their stay in the Olah-group. My office door always was and is open without any formality to any member of my group to discuss chemistry or any personal question or problems they want. Because Judy was an active, key member of our group during much of my academic career, my students got used to unburdening themselves more to the "kinder, gentler" Olah. They were perhaps somewhat awed by my height (6'5" or, by now, probably 6'3"), not always realizing that underneath the somewhat towering appearance, in fact, there always was a rather caring friend. Although I jealously protected my time to concentrate primarily on my research and scholarly work, I always had time for my students and colleagues. I feel this is a natural responsibility of professors and research advisors. After my Nobel prize, as new obligations and pressures were starting to somewhat limit my time, I learned to politely say "no" to many outside invitations. This led me also to decide not

Judy, in Munich with the Huisgens

to take on any more new graduate students so as not to short-change them. My research, however, continued unabated with my postdoctoral fellows. I also maintain close contact with the graduate students in our Institute, particularly those on whose guidance committees I serve, and those of the research group of my colleague and friend Surya Prakash.

I was blessed with an extremely fine group of graduate students and postdoctoral fellows who over the years joined the Olah-group and became my wider scientific family. I have already mentioned many of those of my Cleveland years. In Los Angeles, my graduate students included besides others Douglas Adamson, Mesfin Alemayehu, Robert Aniszfeld, Massoud Arvanaghi, Joseph Bausch, Donald Bellew, Arthur Berrier, Arwed Burrichter, Daniel Donovan, Robert Ellis, Omar Farooq, Morteza Farnia, Jeff Felberg, Alex Fung, Armando Garcia-Luna, Judith Handley, Michael Heagy, Nickolas Hartz, Ludgar Heiliger, Altaf Husain, Wai Man Ip, Pradeep Iyer, Richard Karpeles, Chang-Soo Lee, Eric Marinez, Lena Ohannesian, Alex Oxyzoglou, Golam Rasul, Mark Sassaman, Tatyana Shamma, Joseph Shih, Maurice Stephenson, Qi Wang, Mike Watkins, and Chentao York.

Postdoctoral fellows of my group in Los Angeles were, among others, George Adler, Thorsten Bach, Alessandro Bagno, Mario Barzaghi,

Patrice Batamack, Jeno Bodis, Jorg-Stephan Brunck, Herwig Buch-holz, Imre Bucsi, Jozef Bukala, Steven Butala, Young-Gab Chun, Françoise and Gilbert Clouet, Miklos Czaun, Denis Deffieux, Hans Doggweiler, Ed Edelson, Ronald Edler, Thomas Ernst, Markus Etzkorn, Wolf-Dieter Fessner, Les Field, Dietmar Forstmeyer, Stephan Frohlich, Xiangkai Fu, Helmut George, Mary Jo Gardina, Alain Goeppert, Paul Mwashimba Guyo, Dong-Soo Ha, Mahommed Hachoumy, Sawako Hamanaka, Toshihiko Hashimoto, Nicholas Head, Thomas Heiner, Rainer Herges, Hideaki Horimoto, Jinbo Hu, Ron Hunadi, JePil Hwang, Shin-ichi Inaba, Jens Joschek, Shigenori Kashimura, Takashi Keumi, Schahab Keyaniyan, Douglas Klumpp, Roland Knieler, Kitu Kotian, V. V. Krishnamurthy, Norbert Krass, Ramesh Krishnamurti, Manfred Kroll, Khosrow Laali, Christian Lambert, Koop Lammertsma, Thomas Laube, Jean-Claude Lecoq, Istvan Ledneczki, Stefan Lehnhoff, Xing-ya Li, Qimu Liao, Thomas Mathew, Ashok Mehrotra, Alfred Mertens, Sergio Meth, Grzegorz Mlostoń, Peter Morton, Tadashi Naka-jima, Subhash Narang, Gebhard Neyer, Koji Nakayama, Alex Orlinkov, Rhuichiro Ohnishi, Istvan Palinko, Pablo L. Palomino, Attila Papp, Don Paquin, Farbrizio Pertusati, Maria Teresa Ramos, Pichika Rama-iah, Marc Piteau, Bhushan Chandra Rao, Tarik Rawdah, V. Prakash Reddy, Christophe Rochin, Roger Roussel, Rolf Rupp, Graham Sand-ford, Sabine Schwaiger, Stephan Schweizer, Jurgen Simon, Brij Singh, Gunther Stolp, Hiroshi Suzuki, Hideki Tashiro, Derrick Tabor, Bela Török, Nirupam Trivedi, Akihiko Tsuge, Yashwant Vankar, Julian Vaughan, Stéphane Walspurger, Qunjie Wang, Czaba Weber, Klaus Weber, Thomas Weber, Jurgen Wiedemann, John Wilkinson, An-hsiang Wu, Takehiko Yamato, Elazar Zadoc, and Miljenko Zuanic.

I am very proud of all members of my scientific family. They did very well in their career and hope they have good memories of their stay with me. Over all, in my career in America, I had some 60 graduate students and some 200 postdoctoral or visiting fellows. The fact that I also worked for 8 years in industry and that there was a need in both Cleveland and Los Angeles to build up my field of research from modest beginnings probably explains the higher ratio of postdoctoral fellows to graduate students. However, because, postdoctoral or visiting fellows generally stayed only for 1–2 years, but graduate students stayed for 3–4 years, the composition of my group was usually equally balanced between them.

Being involved in teaching and directing research work for more than half a century, I always felt that mentoring my younger colleagues was an essential part of my responsibilities. In a way, one of the best judgments of any professor is how his students or associates feel about him and also how they fare in their own careers. It is not unlike how parents feel about their children. To observe with satisfaction and pride the achievements of your scientific family, which in a way is also the continuation of your own work, is most rewarding. Because I was never associated with one of the leading universities, having also spent some of my career in industry, my graduate students usually came from more modest schools than those of their peers who have gone to our premier colleges and universities. This is quite understandable and based on sound reasons. Going to Harvard, Stanford, MIT, or Caltech to earn a graduate degree in itself usually guarantees a good chance for a successful career. On the other hand, motivation and desire also count significantly. Over the years, I have seen with great satisfaction many of my students blossom out and become accomplished and productive scientists, even surprising themselves with how far they were able to go. To achieve this, however, there can be no substitute for hard and dedicated work and continued learning while "doing chemistry." A research group is indeed a suitable incubator providing the conditions and challenges for continued development. My role was that of a catalyst and facilitator, who tried to motivate but at the same time keep our joint effort oriented and on target.

I never felt it necessary to put pressure on my students or associates to achieve goals. Drive must come from within and, of course, in the competitive environment of a research group, from friendly competition with your peers. The professor should be careful even to minimize any pressure, because in research we never really know what idea will work out or fail. Because the professor generally gets the lion's share of credit for the achievements of the students, I believe that he/she is also responsible for failures, mistakes, or any other problems arising in his laboratory. In research, we all must live by high ethical and scientific standards, not only because this is the only right way but also because observations and data cannot be fudged or faked. If your work in chemistry is of any interest, others will sooner or later reproduce it, providing the best way to check your results. If your experimental facts are correct, you are, of course, allowed to interpret them your way. Differing interpretations can sometimes result in heated debates or controversies, but in science this eventually is resolved when new facts become known.

There is a fundamental difference between such scientific controversies and what simply can be called "scientific fraud," i.e., deliberate falsification or fudging of data. Sloppy experimental work or data keeping can also lead to questionable or incorrect conclusions, and, although these violate established scientific standards and must be corrected (as they will), they do not necessarily represent deliberate fraud. In all this, the professor has a strict personal responsibility. As he/she is getting most of the recognition for the accomplishment of the research, it is only natural that he/she must also shoulder the responsibility for any mistakes, errors, or even falsifications. It is not acceptable, for example, to say, as is done on occasions, that if in a research group some overeager or misguided student falsified data for nonexistent results that were then published, the unsuspecting professor was only tricked into what later turned out to be embarrassing false claims. If somebody in your research group comes up with a potentially significant new finding, you must make sure before anything else that the data are correct. How many times have we all experienced short-lived expectations, which were, however, soon shown to have some rather trivial reason or explanation. Concerning fraud, there is little that does not get noticed by one's immediate lab colleagues. A professor, I believe, must have always good contact with his group and be sensitive to pick up "early warning signs" if there are any reasons for concern, long before publication. In the case of results that may indicate a breakthrough of potential significance, care is, of course, even more important. In this case, it is customary to repeat the work several times over, generally also by other members of the group. Checking data and avoiding any cheating or fudging must be self-controlled in the laboratory. Moreover, if you do not catch the mistake yourself, others will do it for you. Blaming a student or associate publicly for whatever mistakes may have been made seems equally wrong to me. This can be an internal "family matter" but is no excuse. As a parent, you are responsible for your children living in your house, and you have to better know what they are doing. The same is true for your laboratory.

In the nonclassical ion controversy discussed in Chapter 9, there was never any question on either side of the debate about the validity of the observed data, only about their interpretation. Had any of the experimental data been questioned or found to befalsified, this would have been soon found out because so many people repeated and rechecked the data. This is the strength of science (in contrast to politics, economics, etc.), i.e., that we deal with reproducible experimental observation and data. Nevertheless, interpretation can still result in heated discussions or controversies, but science eventually will sort these out based on repeated and new results and data.

The final evaluation and judgment of any research and its significance come from the wider scientific community upon its publication. In my research, I have always believed that there is a fundamental obligation to publish our work in leading peer-reviewed journals. This allows not only communication of the work to the wider scientific community but also its evaluation and, if needed, criticism by others around the world. No research is complete without publication of its results and conclusions. I also felt a strong obligation toward my collaborators and students, who really made the research possible and contributed fundamentally to it. Even if after a time your own drive to see your name on publications (which as in case of any other intellectual or artistic activities certainly is a factor) may be diminished, your younger colleagues, who spent some of their best years on the research project, deserve to see their work published to receive full credit for it. Publishing scientific work is clearly an essential part of the research process. If one is not interested or prepared enough to write up and publish the results of the work in a relatively timely manner, valid questions can be raised as to whether the research should have been undertaken at all. Some criticize researchers whose publications they believe are too numerous. I may be one of these, because I have published some 1500 research papers (in refereed journals, excluding other printed materials such as abstracts of lectures, reviews, comments, and letters). My goal, however, was never to "pad" my publication list. I myself never paid much attention to the number of my publications, although some of my friends and colleagues occasionally did comment on it. As I said earlier, I always felt and still feel that publishing is an integral part of any research or scholarly project and should be considered in this context.

Chemistry is also a practical science with a very significant industrial base. University education in chemistry including graduate schools, however, hardly prepares one for a career in industrial chemistry. Although chemistry faculties generally are composed of very competent and capable chemists providing a thorough chemical education, most lack industrial experience that they could convey to their students. I was fortunate in this regard that I had spent 8 years in industry and thus had a realistic first-hand view to present to my students about both academic and industrial chemistry. My industrial experience, I feel, served me well also as a teacher and a mentor for my students who were going into an industrial career.

Another way in which academic chemists keep in touch with industry is through consulting. During my Cleveland years I consulted at different times for Exxon, Chevron, Cyanamid, ENICHEM (Italy), and Pechiney-Kuhlman (France). I enjoyed these contacts, because they kept me aware of current industrial developments and interests. At the same time, it was also rewarding to be able to suggest to my industrial friends new approaches and directions that on occasion were useful and resulted in practical applications.

A more limited aspect of my consulting over the years involved some legal cases connected with my fields of expertise. The most interesting of these involved the Ziegler–Natta polymerization of olefins. It involved the work of Max Fischer, a German industrial chemist during World War II, in the German chemical industry who polymerized ethylene with $AlCl_3$ and Al powder in the presence of $TiCl_4$ under moderate pressure and obtained polyethylene. In the extensive litigation of the Ziegler–Natta polymerization the question arose of whether Fischer's work was a Friedel–Crafts type ionic polymerization or a forerunner of the Ziegler polymerization. I found the question so fascinating that subsequently we carried out some fundamental research into the underlying chemistry of $AlCl_3/Al$, which can form aluminum dichloride and related organoaluminum compounds.

I gave up most consulting after moving to Los Angeles and starting the Loker Hydrocarbon Research Institute. To build the Institute and to assure its future became my priority, and I felt that I should not dilute my effort with much private consulting. Our Institute concentrated from the beginning its efforts on fundamental research and graduate training in the broad area of hydrocarbon chemistry. We decided not to accept any contract work either or support from industry that would directly involve working on their specific problems of the donors. Instead, when our chemistry resulted in discoveries and we felt that they may have practical use, we patented them. The Institute subsequently welcomed arrangements with the interested industry to acquire and further develop our chemistry for practical uses.

Starting with my work in Hungary, through my years with Dow Chemical, and during my academic career, besides scientific publications, I thus applied for patents on discoveries I felt represented practical significance. Over the years I obtained close to 170 patents including still pending applications. In the Loker Institute, patents and intellectual property generated by our work are assigned to our university and contribute some support for the research of the Institute.

Universities or university-related institutes, except in such fields as computer and software technology as well as biotechnology among others, are still learning how to utilize the technical discoveries and know-how of their faculty. In our field of hydrocarbon research this is a particularly difficult challenge. No one can start an oil refining or petrochemical operation in his garage or backyard. Cooperation with industrial organizations with their technological expertise and ability to work toward commercialization can be mutually advantageous.

In the last 15 years, with the help of my friends and colleagues Surya Prakash and Thomas Mathew, we completed in three volumes and published a collection of my selected papers and patents together with synopsis and comments organized according to the topics of my interest covered (*Across Conventional Lines* published by World Scientific Publishing Co., Singapore, also the Nobel Foundations Publisher) which include a complete bibliography (see Appendix for intended readers).

As discussed in Chapters 13–15, after my Nobel Prize of 1994, I concentrated mainly on the goal to find a practical and economic solution to replace petroleum oil which is getting scarce and less readily available, clearly a major challenge for the future. The work resulted in developing what is now called the "Methanol Economy" and starting to get practical appreciation around the world.

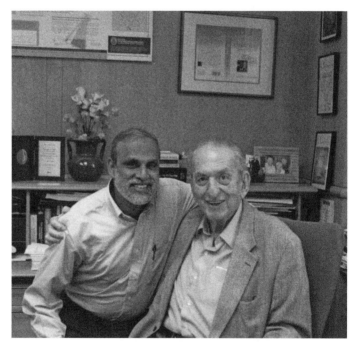

With Thomas Mathew

My work in superacid and related synthetic chemistry also developed a series of useful reagents. In the early 1970s, I helped a former graduate student of mine, Jim Svoboda, to start a small company in Cleveland called Cationics, to make and sell our cationic reagents and superacidic systems that were not readily available to the wider chemical community. He and his wife (with Judy's initial help) made a great effort, and within a short time the small company gained some recognition and helped to make our reagents better known and available. They were eventually (and still are) distributed by some of the major fine chemical and reagent companies (Aldrich, Alpha, among others). Financially, however, Cationics was hardly viable. Jim learned the hard way that money is not necessarily made in the field by laboriously preparing reagents but by selling them with a substantial profit margin and obtaining them from other sources (including struggling small companies). Cationics moved from Cleveland to Columbia, South Carolina, was subsequently absorbed into Max Gergel's Columbia Chemicals, and eventually faded away. None of us made any money on it, despite the significant effort expended on it. For me, it was a learning experience not to venture to start a chemical start-up company if you are and want to stay a scientist and not a businessman.

In my later Cleveland years, I also gained some other experience in the emerging area of high-technology companies by being asked to join the board of directors of a company involved to commercialize a new type of disposable chemical personal thermometer replacing mercury. It was based on the color change of a series of separate dots containing eutectic mixtures of two suitable chemicals melting $0.1°C$ apart and encapsulated in a small plastic strip to be placed under the tongue. The idea of a disposable chemical thermometer was quite ingenious, and other applications could be foreseen, including an indicator strip registering the maximum temperature that packaged frozen foods were exposed to during storage. The company (Bio-Medical Sciences, BMS) was well financed (by some $25 million at the time impressive for a small startup), and its investors included the Prudential Insurance Company, the Rockefeller Brothers Trust, and Yale University. Replacement of the mercury thermometer also with other devices (including electronic ones) subsequently came about. BMS itself was bought out and its product is still under wide uses. In any case, when I moved to Los Angeles I gave up any outside interests, because I felt it proper to concentrate only on my efforts at USC.

Life presents us all with many challenges. I learned early in my life to face such challenges, including just for survival in the difficult and troubled World War II years in my native Hungary. The subsequent years in a poor and much destroyed country moving from one extreme regime to another were also not easy. It is said that hardship shapes your character and strengthens your spine, but after a while you feel that perhaps you need no more of it. Thus, when in 1956 my family and small research group escaped a rather dismal situation of living in a Soviet dominated communist state and we started a new life in America, I was rather well prepared to cope with life's challenges. I am most grateful to the United States my new adopted country, which gave me the opportunity to restart my career in chemistry and provided me and my family a new home. When I was asked years later to write a brief statement at the end of my Who's Who biographical sketch, I wrote, "America is still offering a new home and nearly unlimited possibilities to the newcomer who is willing to work hard for it. It is also where the 'main action' in science and technology remains." I still feel the same way.

We also frequently face other challenges in life. Some 30 years ago I went through two life-threatening health crises involving a stomach and rare acute genetic disease problem. I came through and I was able to continue my work with little interruption. You learn from such "close encounters" what is really important in your life. I believe I owe much to having learned from hardships and challenges, and I appreciate the love and support of my family more than I can express. In 2005, I underwent two knee replacements and in 2010 a partial hip replacement and recovered well. Recently in early 2014, I was diagnosed with bladder cancer and had undergone extensive surgery. I am fortunate again to recover without apparently the need of further treatment. I am able to again actively continue work including writing and finishing books (such as this one) and my research effort mainly toward the further development of the "Methanol Economy."

I am very grateful for the continuous support of the University of Southern California over the years and in particular its presidents including John R. Hubbard, James H. Zumberge, and Steven B. Sample.

Professor C. L. Max Nikias, the present president continues this most gratifying support.

I much appreciate the privilege of having been able to work with wonderful students and colleagues. I am also grateful that I was able to follow my own ways and principles, with all the shortcomings I have. From the adversities I faced and difficulties I encountered, I learned to look only forward and to make the best out of my possibilities. I never regretted or second-guessed the pathway of my life. Life anyhow is too short to worry much, and it is better to concentrate on moving forward. I also learned to better distance myself from the many inevitable upsets and irritations we all face, but which after a while turn out to be quite unimportant. In your professional work, too, you learn to differentiate what is superficial from what is really meaningful. I also learned to politely say "no" to many invitations, involvements, etc., which however worthwhile, inevitably would distract from pursuing my essential goals.

I continue my best efforts to contribute to the growth and the excellence of USC. I am staying active and with all modesty still quite productive. Together with Judy, we also continue our support of our University. We established a fellowship for deserving undergraduate students and continue support for various activities of our Hydrocarbon Research Institute. I have assigned all my intellectual property (patents) rights to USC and actively help in efforts to bring to practical use of some of the 40 patents we obtained together with my colleague Surya Prakash in the area of the "Methanol Economy."

In 2010, I turned over the directorship of the Loker Hydrocarbon Research Institute together with all administrative duties to my friend Surya Prakash. At the request of President Nikias and Provost Garret, I continue, however, as distinguished professor of Chemistry and Engineering and also became the chairman of Institute's advisory board following Mrs. Nicole Keck who took over after the death of Mrs. Loker but regrettably could not continue. Our advisory board was reorganized with valuable new members such as Messrs Chrystopher Cox, Carlos Cabrera, and Alexander L. Cappello as well as on the scientific side, the participation of Peter Stang, Gabor Somorjai, and recently Arieh Warshel.

Scientists or, for that matter, writers, artists, performers, and athletes, etc., cannot deny that acknowledgment of their accomplishments is satisfying and rewarding. During a long career in science, you receive different prizes, medals, and recognitions of your work. I received from the American Chemical Society my first award in 1964 in Petroleum Chemistry (which was later endowed and renamed as the "George A. Olah Award in Hydrocarbons or Petroleum Chemistry"). In 1979, I won the Award for Creative Work in Organic Synthesis; in 1989, the Roger Adams Award; in 2001, Arthur C. Cope Award; and in 2005, the Priestly Medal, the American Chemical Society's most prestigious award. Previously, I also was given the Baekeland and Tolman Awards and Morley Medal of ACS. Other recognitions included two Guggenheim Fellowships and a senior Humboldt Award. Japan honored me with the "Medal of the Rising Star," the highest award for a foreigner. My native Hungary also bestowed on me various state awards and medals. I was also awarded the Széchenyi Grand Prize of Hungary in 2011 and the Semmelweis Budapest Award in 2013. In 2013, the inaugural Eric & Sheila Samson Prize by the Prime Minister of Israel for innovation in alternative fuels and transportation was awarded to me together with Professor Prakash, which I particularly appreciated as a recognition of our ongoing current efforts in the area of renewable energy and the development of the practical and feasible "Methanol Economy" concept. One of the most pleasing recognitions I ever received was the Cotton Medal in 1996, named after and presided by Al Cotton, one of the leading inorganic chemist in the world, an outspoken advocate of science and education, and a close late friend. Receiving recognitions from your peers, who can judge your work best, is always rewarding.

I was unexpectedly honored by some frescos dedicated to me and painted by Peter and Pavel Mester which were selected and commissioned at the initiative of the late Nobel Prize winner Sir Andrew Fielding Huxley and adorn the Chemistry—Engineering Center of Cambridge University in England. Later some of related to these were also painted for our Loker Hydrocarbon Research Institute at USC. They are part of a larger collection of paintings at various Universities and Institutes to further closer ties between scientists and artists leading to artwork inspired by science.

Cotton Medal from Al

Wall painting by Peter and Pavel Mester installed at the Loker Hydrocarbon Institute

It is necessary to keep recognitions in proper perspective. Receiving an award or prize (even the Nobel Prize) is based on subjective judgment and no judgment or evaluation can be perfect. Life goes on, and it is frequently more revealing to observe how people continue in their field after public recognitions. I am myself careful to differentiate some recognitions I received before my Nobel Prize from those which were given me afterwards. You tend, for example, to collect honorary degrees, memberships in learned societies and academies, etc., for which you probably would not have been considered otherwise (see my curriculum vitae in the Appendix). In any case, during my career I have been elected to more than a dozen of scientific academies as well as honorary member to varied societies which I am proud of. These include inter alia the US National Academy of Sciences, the British Royal Society, the Italian National Academy-Lincei, the Canadian Royal Society, and the Hungarian Academy of Sciences. Election to such bodies is certainly an honor, although once elected you find out that members generally take little further role in their activities.

I also received some 15 honorary degrees, the first from the University of Durham, in 1988 presented by its Vice-Chancellor of the time Dame Margot Fonteyn, the famous ballet dancer and a remarkable lady. Others came from the University of Munich (1990), University of Crete (1994), my alma mater the Technical University of Budapest (1989), the Universities of Szeged and Veszprem in Hungary (1995), University of Montpellier (1996), New York State University (1998), University of Pecs in Hungary (2001); University of Debrecen in Hungary (2003), University of Western Hungary (2006), Ohio State University (2009), and Obuda University in Hungary, 2011.

Regrettably, our life, including science and higher education, is becoming increasingly bureaucratic. Talented researchers spend much of their time in committees, boards, and panels instead of pursuing their research. I have always tried to keep centered on my work and not sidetracked, remembering a short story by Leo Szilard, the remarkable Hungarian-born scientist, included in his book *The Voice of the Dolphins*, published in the 1950s. It tells of a scientist who had his body frozen to come back years later. He then became a celebrity in his city and on occasion met a very wealthy man who asked him what he should do with his fortune. The scientist suggested that he should set up a foundation supporting scientific work. The millionaire answered that he hardly would consider it, because he intensely disliked science. After some consideration our scientist, however, came up with a solution. You should still establish your foundation, he advised, but insist that its board as well as its numerous committees should be composed of the most productive and promising scientists. To assure this, the foundation would pay them such high honoraria that nobody would decline. With all these outstanding scientists fully involved in their bureaucratic assignments, their productivity will rapidly decrease and eventually cease. Science will thus wither, and you will achieve your goal. There is much to think about in this imagined story. In fact, too many scientists are becoming increasingly less productive by allowing themselves to be overwhelmed by many commitments outside their research. In some areas such as the computer-electronics and biotechnology fields, commercial interests also tend these days to overshadow science. I myself always tried to avoid such things as much as possible and stayed focused. Others may say that avoiding committees, etc., is a very selfish attitude, but are creative artists, writers, and composers any less selfish?

I have always tried to keep a healthy sense of humor, much needed in our present time. Similarly, I have managed not to take myself too seriously, only my science, about which I am quite passionate. This is useful to emphasize, because for scientists in the long run what really counts is what they have achieved, not their personality or ego. To stay productive and may be even creative, one must maintain curiosity and enjoy the pleasure of being able to search for new understanding and of making occasional discoveries and contributions may be of some importance (if they still come their way). Those who follow success and recognition frequently succumb to an overblown feeling of self-importance, frequently missing out on these. It is sometimes said that Nobelists who receive their prize later in life are lucky, because after the prize one's scientific productivity generally decreases. This indeed may be the case generally, but it is possible to avoid it. I considered myself one of the lucky ones who succeeded in avoiding this trap. I believe that the key is how well you can focus on continuing your scientific work unabated, if this is really your goal. To keep my priorities focused and to be able consistently to say *no* to the many worthwhile activities the prize inevitably presents are the most important guidelines I learned and they served me well.

· 17 ·

Concluding Remarks

My long journey which started in my native Budapest on the banks of the river Danube and took me to the shores of the Pacific Ocean was not always an easy one. Human nature, however, helps to block out memories of hardship and difficulties. They fade away and you look back remembering mostly the positive aspects of your life. I followed my own principles and went my own way. It helped that I inherited a strong, perhaps on occasion, stubborn nature with a determination to follow the pathway to my goals and that I worked hard to achieve them.

It was a life full with challenges and never diminishing devotion to chemistry. My extended greater family of generations of students, colleagues, and associates was and still is an integral part of my life. Jointly, we explored the ever-changing and developing world of chemistry and may have contributed in our way to its continuing development. It is also of great satisfaction that in the last decades our work resulted even in practical advances, contributing for example to feasible new solutions to the replacement of our diminishing fossil fuels and environmentally adaptable cleaner ways to solve our carbon conundrum.

I am still involved in writing and planning books with coopera-tion of my colleagues particularly Surya Prakash, Alain Goeppert, and Thomas Mathew. These include a monograph on single carbon *C1 Chemistry* (with Thomas Mathew and Surya Prakash, Wiley-VCH), a popular book on the future of safe and lasting use of carbon-containing fuels/materials and products, a new updated third edition

A Life of Magic Chemistry: Autobiographical Reflections Including Post-Nobel Prize Years and the Methanol Economy, Second Updated Edition. George A. Olah with Thomas Mathew.
© 2015 John Wiley & Sons, Inc. Published 2015 by John Wiley & Sons, Inc.

of our book with Alain Goeppert and Surya Prakash on the *Methanol Economy* (Wiley-VCH) and extensive publications of our research as well as reviews and varied commentaries. My colleague, Professor Surya Prakash, is increasingly taking over our joint research areas and their continued future developments. Effective further practical applications will ensure the bright future of our Institute's work in years to come. I could leave the continuation of my work in no better hands.

The most important blessing of my life is my family. Looking back at the twilight of my life, I am blessed with the best wife everybody can imagine and a wonderful family of my two sons George and Ron, our two daughters-in-law, Sally and Cindy and three grandchildren, Peter, a junior at USC, Kaitlyn, a freshman at Stanford and Justin, who is starting high school next fall.

Judy, my wife of 65 years, was and is the pillar of my life and our family. She also contributed to and made possible my scientific work to flourish. We even shared our professional work and she was also the cement which held together my scientific family. I cannot express my gratitude for the love and strength she always gave me, although having deprived her of much valuable time which we could have spent by leisurely just being together. I wish that I have changed much and

The Olah family (December 2013)

spent much more time with her and my family. I am glad that our two sons, George and Ron, have not followed us into the area of science. George, an MBA graduate, is a partner and chief financial officer of a small financial company. Ron, a physician, is practicing internal medicine in close-by Pasadena. They both have their family and successful careers. I was blessed by always having had their help and support. I ask again for Judy's and their forgiveness that my scientific work took away so much time I could have spent with them. They gave however, all that a family can give with love and understanding. What else could I ever have asked for?

· Appendix ·

My Previous Books for References and Additional Reading

Introduction to Theoretical Organic Chemistry (in German), Akademie Verlag, 1960.

Friedel-Crafts and Related Reactions (ed.), Vols. I–IV, Wiley-Interscience, 1963–1965.

Carbonium Ions (ed. with Paul v. R. Schleyer), Vols. I–V, Wiley-Interscience, 1968–1973.

Carbocations and Electrophilic Reactions (German and English versions), Verlag Chemie-Wiley-Interscience, 1973.

Friedel-Crafts Chemistry, Wiley-Interscience, 1973.

Halonium Ions, Wiley-Interscience, 1975.

Superacids (with G. K. S. Prakash and J. Sommer), Wiley-Interscience, New York, 1985.

Hypercarbon Chemistry (with G. K. S. Prakash, K. Wade, L. D. Field, and R. E. Williams), Wiley-Interscience, New York, 1987.

Nitration: Methods and Mechanisms (with R. Malhotra and S. C. Narang), VCH Publishers, New York, 1989.

Cage Hydrocarbons (ed.), Wiley-Interscience, New York, 1990.

A Life of Magic Chemistry: Autobiographical Reflections Including Post-Nobel Prize Years and the Methanol Economy, Second Updated Edition. George A. Olah with Thomas Mathew.
© 2015 John Wiley & Sons, Inc. Published 2015 by John Wiley & Sons, Inc.

Electron Deficient Boron and Carbon Clusters (ed. with R. E. Williams and K. Wade), Wiley-Interscience, New York, 1991.

Chemistry of Energetic Materials (ed. with D. R. Squire), Academic Press, New York, 1991.

Synthetic Fluorine Chemistry (ed. with R. D. Chambers and G. K. S. Prakash), Wiley-Interscience, New York, 1992.

Hydrocarbon Chemistry (with Á. Molnár), Wiley-Interscience, New York, 1994, 2nd Revised Edition, 2003.

Onium Ions (with K. K. Laali, Q. Wang, and G. K. S. Prakash), Wiley-Interscience, New York, 1998.

A Life of Magic Chemistry: Autobiographical Reflections, John Wiley & Sons, New York, 2000.

Across Conventional Lines (Selected papers with commentary) (ed. with G. K. S. Prakash), Vols. 1–2, World Scientific Publisher, Singapore, 2003.

Carbocation Chemistry (with G. K. S. Prakash), John Wiley & Sons, New York, 2004.

Beyond Oil and Gas: The Methanol Economy (with A. Goeppert and G. K. S. Prakash), Wiley-VCH, 2006 (also Chinese, Hungarian, Japanese, Russian, and Swedish editions, 2007).

Superelectrophiles and Their Chemistry (with D. A. Klumpp), John Wiley & Sons, 2008.

Superacid Chemistry (with G. K. S. Prakash, Á. Molnár, and J. Sommer), John Wiley & Sons, 2009.

Beyond Oil and Gas: The Methanol Economy (Second Updated and Enlarged Edition with A. Goeppert and G. K. S. Prakash), Wiley-VCH, 2009 (also Chinese and Japanese editions, 2010).

La ricerca di una vita, Di Renzo Editore, Rome, 2010.

Hypercarbon Chemistry (Enlarged Second Edition with G. K. S. Prakash, R. Williams, K. Wade, and Á. Molnár), John Wiley & Sons, 2011.

Across Conventional Lines, Vol. 3. The Sixth Decade and the Methanol Economy (ed. with G. K. S. Prakash and Thomas Mathew), World Scientific Publishing Co., Singapore, 2014.

My work was published in some 1500 peer reviewed scientific articles with details and references. I have some 170 patents. The interested reader can find a complete bibliography in the collection of my selected papers Across Conventional Lines, World Scientific Publisher, Singapore, 2014.

Curriculum Vitae of
George Andrew Olah

Born (May 22, 1927) and educated in Budapest, Hungary. M.S. in Chemical Engineering and Ph.D., Technical University Budapest, Hungary, 1949, under the late Professor Geza Zemplen (Organic Chemistry). Naturalized U.S. Citizen, 1971. Married Judith Agnes Lengyel, 1949. Children, 2 (George, 1954 and Ronald, 1959).

Professional Experience

1949–1954 Assistant Professor of Organic Chemistry, Technical University, Budapest, Hungary

1954–1956 Head of Department of Organic Chemistry and Associate Scientific Director of Central Research Institute of the Hungarian Academy of Sciences, Budapest, Hungary

1957–1964 Research Scientist, Dow Chemical Company

1965–1967 Professor and Chairman, Department of Chemistry, Western Reserve University, Cleveland, Ohio

1967–1969 Chairman of combined departments of chemistry of Case Institute of Technology and Western Reserve University

A Life of Magic Chemistry: Autobiographical Reflections Including Post-Nobel Prize Years and the Methanol Economy, Second Updated Edition. George A. Olah with Thomas Mathew.
© 2015 John Wiley & Sons, Inc. Published 2015 by John Wiley & Sons, Inc.

1967–1977 C. F. Mabery Distinguished Professor of Research in Chemistry, Case Western Reserve University

1977 Professor of Chemistry and Scientific Director, Hydrocarbon Research Institute, University of Southern California

1980– Distinguished Donald P. and Katherine B. Loker Professor of Chemistry, University of Southern California

1991–2010 Director, Loker Hydrocarbon Research Institute, University of Southern California

2008– Distinguished Professor, School of Engineering, University of Southern California

2010– Founding Director, Loker Hydrocarbon Research Institute, University of Southern California

Visiting and Honorary Professorships

Various times visiting professor at Ohio State University; University of Heidelberg; University of Colorado; Swiss Federal Institute of Technology; University of Munich; University of Strasbourg; University of Montpellier; University of London, King's College; University of Paris; Honorary Professor. University of Dubna, Russia, 2000–; Institute of Organic Chemistry, Chinese Academy of Sciences, Shanghai, 2003–; Peking University, 2007.

Publications and Patents

Published more than 1500 scientific papers and holds 170 patents. Books include: *Introduction to Theoretical Organic Chemistry* (in German), Akademie Verlag, 1960; *Friedel-Crafts and Related Reactions* (ed.), Vols. I–IV, Wiley-Interscience, 1963–1965; *Carbonium Ions* (ed. with Paul v. R. Schleyer), Vols. I–V, Wiley-Interscience, 1968–1973; *Carbocations and Electrophilic Reactions* (German and English versions), Verlag Chemie-Wiley-Interscience, 1973; *Friedel-Crafts Chemistry*, Wiley-Interscience, 1973; *Halonium Ions*, Wiley-Interscience, 1975; *Superacids* (ed. with G. K. S. Prakash and J. Sommer), Wiley-Interscience, New York, 1985; *Hypercarbon Chemistry* (ed. with

G. K. S. Prakash, K. Wade, L. D. Field, and R. E. Williams), Wiley-Interscience, New York, 1987; *Nitration: Methods and Mechanisms* (ed. with R. Malhotra and S. C. Narang), VCH Publishers, New York, 1989; *Cage Hydrocarbons* (ed.), Wiley-Interscience, New York, 1990; *Electron Deficient Boron and Carbon Clusters* (ed. with R. E. Williams and K. Wade), Wiley-Interscience, New York, 1991; *Chemistry of Energetic Materials* (ed. with D. R. Squire), Academic Press, New York, 1991; *Synthetic Fluorine Chemistry* (ed. with R. D. Chambers and G. K. S. Prakash), Wiley-Interscience, New York, 1992; *Hydrocarbon Chemistry* (ed. with Á. Molnár), Wiley-Interscience, New York, 1994, 2nd Revised Edition, 2003; *Onium Ions* (ed. with K. K. Laali, Q. Wang, and G. K. S. Prakash), Wiley-Interscience, New York, 1998; *A Life of Magic Chemistry: Autobiographical Reflections*, John Wiley & Sons, New York, 2000; *Across Conventional Lines* (Selected papers with commentary) (ed. with G. K. S. Prakash), World Scientific Publisher, Singapore, 2003; *Carbocation Chemistry* (ed. with G. K. S. Prakash), Wiley-Interscience, New York, 2004; *Beyond Oil and Gas: The Methanol Economy* (ed. with A. Goeppert and G. K. S. Prakash), Wiley-VCH, 2006 (also Chinese, Hungarian, Japanese, Russian, and Swedish Editions, 2007); *Superelectrophiles and Their Chemistry* (with D. Klumpp), John Wiley & Sons, 2008; *Superacid Chemistry* (with G. K. S. Prakash, Á. Molnár, and J. Sommer), John Wiley & Sons, 2009; *Beyond Oil and Gas: The Methanol Economy* (Second Updated and Enlarged Edition with A. Goeppert and G. K. S. Prakash), Wiley-VCH, 2009 (also Chinese and Japanese Editions, 2010); *Hypercarbon Chemistry* (2nd Edition with G. K. S. Prakash, R. Williams, K. Wade, and Á. Molnár), John Wiley & Sons, 2010; *La ricerca di una vita*, Di Renzo Ed., Rome, 2010; *Across Conventional Lines: The Sixth Decade and the Methanol Economy* (Selected papers with commentary) (ed. with G. K. S. Prakash and Thomas Mathew), World Scientific Publishing Co., Singapore, 2014.

Awards

American Chemical Society Award in Petroleum Chemistry, 1964

Leo Hendrick Baekeland Award, American Chemical Society, 1967

Morley Medal, American Chemical Society, 1970

Fellow of the J. S. Guggenheim Foundation, 1972 and 1988

Fellow of the Society for the Promotion of Science, Japan, 1974

Member of the U.S. National Academy of Sciences, 1976

Centenary Lectureship, British Chemical Society, 1977

American Chemical Society Award for Creative Work in Synthetic Organic Chemistry, 1979

Alexander von Humbolt-Stiftung Award for Senior U.S. Scientists, 1979

Foreign Member of the Italian National Academy dei Lincei, 1982

Michelson-Morley Award of Case Western Reserve University, 1987

Honorary Member, Italian Chemical Society, 1988

California Scientist of the Year Award, 1989

American Chemical Society Roger Adams Award in Organic Chemistry, 1989

Member of the European Academy of Arts, Sciences and Humanities, 1989

Honorary Member Hungarian Academy of Sciences, 1990

Richard C. Tolman Award, American Chemical Society, Southern California Section, 1992

Chemical Pioneers Award, The American Institute of Chemists, Inc., 1993

William Lloyd Evans Award, The Ohio State University, 1993

Nobel Prize in Chemistry, 1994 (unshared)

George Washington Award, American Hungarian Foundation, 1995

Cotton Medal, American Chemical Society, Texas A&M, 1996

Foreign Member, Russian Academy of Natural Sciences, 1996

Member, American Academy of Achievement, 1996

Kapitsa-Medal Russian Academy of Natural Sciences, 1996

Inventor of the Year Award, New York Intellectual Property Lawyer Association, 1996

The American Chemical Society renamed its annual Award in Petroleum Chemistry as the "George A. Olah Award in Hydrocarbon or Petroleum Chemistry," 1996

Golden Plate Award, American Academy of Achievement, 1996

Foreign Member of the Royal Society of London, 1997

Foreign Member of the Royal Society of Canada, 1997

Government of Hungary Award "The Promotion of the Renown of Hungary," 1997

Gold Medal of Charles University, Prague, Czech Republic, 1999

Hanus Medal, Czech Chemical Society, 1999

Honorary Fellow, Royal Society of Chemistry (London), 1999

Arthur C. Cope Award, American Chemical Society, 2001

Order of the Corvin Chain of Hungary, 2001

Albert Einstein Medal, Russian Academy of Natural Sciences, 2002

Honorary Member, Chemical Society of Japan, 2002

Member of American Academy of Arts and Sciences, 2002

Member, American Philosophical Society, 2002

Bolyai Prize, Hungarian Academy of Science, 2002

Member of Royal Academy of Sciences and Arts, Barcelona, Spain, 2003

Klebelsberg Prize of the University of Szeged, Hungary, 2003

Grand Cordon of the Order of the Rising Sun (Japan), 2003

Fellow of the European Academy of Sciences, 2004

Crown & Eagle Medal, Russian Academy of Natural Sciences, 2004

American Chemical Society Priestley Medal, 2005

Hungarian Order of Pro Merit, 2006

Honorary Citizen of the City of Budapest, Hungary, 2007

Honorary Member, German Chemical Society, 2007

Member, U.S. National Academy of Engineering, 2009

Honorary Fellow of the American Chemical Society, 2009

Széchenyi Grand Prize of Hungary, 2011

Semmelweis Budapest Award, 2013

Israel Prime Minister's Prize for Innovation in Alternative Fuels for Transportation, 2013

Honorary Degrees

D.Sc. Honoris Causa: University of Durham, England, 1988; Technical University of Budapest, 1989; University of Munich, 1990; University of Crete, 1994; University of Jozsef Attila, Szeged, Hungary, 1995; University of Veszprém, Veszprém, Hungary, 1995; University of Southern California, Los Angeles, 1995; Case Western Reserve University, Cleveland, Ohio, 1995; University of Montpellier, Montpellier, France, 1996; State University of New York, 1998; University of Pecs, Hungary, 2001; University of Debrecen, Hungary, 2003; University of Western Hungary, 2006; Ohio State University, 2009; Obuda University, Hungary, 2011.

Index

A Life of Magic Chemistry: Autobiographical Reflections Including Post-Nobel Prize Years and the Methanol Economy, Second Updated Edition. George A. Olah with Thomas Mathew.
© 2015 John Wiley & Sons, Inc. Published 2015 by John Wiley & Sons, Inc.